9th International Conference "Modern Nanotechnologies and Nanophotonics for Science and Industry"

Journal of Physics: Conference Series Volume 2316

Vladimir, Russia
11-16 November 2021

ISBN: 978-1-7138-6037-2
ISSN: 1742-6588

Printed from e-media with permission by:

Curran Associates, Inc.
57 Morehouse Lane
Red Hook, NY 12571

Some format issues inherent in the e-media version may also appear in this print version.

Institute of Physics
Dirac House, Temple Back
Bristol BS1 6BE UK

Phone: 44 1 17 929 7481
Fax: 44 1 17 920 0979

techtracking@iop.org

Additional copies of this publication are available from:

Curran Associates, Inc.
57 Morehouse Lane
Red Hook, NY 12571 USA
Phone: 845-758-0400
Fax: 845-758-2633
Email: curran@proceedings.com
Web: www.proceedings.com

TABLE OF CONTENTS

The Vladimir State University special issue based on the results of the conference MNNSI-2021

The main goal of the conference: Getting together of young scientists (Master-students, PhD students, Postdocs) and leading researchers for a joint discussion of actual research problems of modern fundamental and applied physics, modalities of training by research.

Organizers

- Stoletovs' Vladimir State University, Vladimir, Russia,
- Westlake University of Hangzhou, China,
- Russian Quantum Center, Skolkovo, Russia.

Supporters

- Ministry of Science and Higher Education of the Russian Federation,
- NT-MDT Spectrum Instruments,
- Avesta Project Ltd.,
- OSTEC Group,
- Russian Quantum Center.

Conference Chairman

Sergey Arakelian – Head of the Department of Physics and Applied Mathematics Vladimir State University, Dr.Sc., Prof.

Chairman of the Organizing Committee

Stella Kutrovskaya – Leading researcher of the Department of Physics and Applied Mathematics Vladimir State University, Dr., Associate professor.

Program Committee

Alexey Kucherik – Head of department of research and information technology of Vladimir State University, Dr.Hab., Associate professor, Chairman

Anton Osipov – Senior researcher of the Department of Physics and Applied Mathematics Vladimir State University, PhD, Scientific Secretary.

Andrey Evlyukhin – Institute of quantum optics, Leibniz Universität Hannover, Germany, Dr.Sc., PhD.

Sergey Garnov – Director of General Physics Institute RAS, corresponding member of RAS, Dr.Sc., Prof.

Sergey Kilin – Deputy Chairman of the Presidium of the Belarusian National Academy of Sciences, academician, Dr.Sc., Prof.

Tatiana Itina – Université de Lyon, Laboratoire Hubert Curien, Saint-Etienne, France, Dr.Sc., Prof.

Alexey Kavokin – Director of International Centre for Polaritonics, Westlake University, Hangzhou, China, Ph.D., Prof.

Vladimir Makarov – Head of the Department of General Physics and Wave Processes, Director of the ILC MSU. Dr.Sc., Prof.

Igor Ryabtsev – Head of the Laboratory of nonlinear resonant processes and laser diagnostics ISP SB RAS, corresponding member of RAS, Dr.Sc., Prof.

Alexey Taichenachev – Director of ILP SB RAS, corresponding member of RAS, Dr.Sc., Prof.

Alexander Shkurinov – Director of ILIT RAS, corresponding member of RAS, Dr.Sc., Prof.

Tigran Vartanyan – Chief Researcher, University ITMO, Dr.Sc., Prof.

Vadim Veiko – Head of International Laboratory "Laser Micro-and Nanotechnologies, University ITMO, Dr.Sc., Prof.

Viktor Zadkov – Director of the Institute of Spectroscopy, RAS, Dr.Sc., Prof.

Peer Review Statement

All papers published in this volume have been reviewed through processes administered by the Editors. Reviews were conducted by expert referees to the professional and scientific standards expected of a proceedings journal published by IOP Publishing Publishing.

- **Type of peer review:** Double Anonymous
- **Conference submission management system:** Morressier
- **Number of submissions received:** 20
- **Number of submissions sent for review:** 20
- **Number of submissions accepted:** 20
- **Acceptance Rate (Submissions Accepted / Submissions Received × 100):** 100
- **Average number of reviews per paper:** 1.2
- **Total number of reviewers involved:** 2
- **Contact person for queries:**
  Name: Anton Osipov
  Email: Osipov@vlsu.ru
  Affiliation: Vladimir State University

Journal of Physics: Conference Series **2316** (2022) 012001 doi:10.1088/1742-6596/2316/1/012001

AI Acceleration Enabled by Nanoelectronic Memristive Devices

I A Bordanov[1], A D Zuev[1], S N Danilin[1], V P Khranilov[2], E A Bukvarev[2], S Kim[3], E G Gryaznov[4], A N Mikhaylov[4], S A Shchanikov[1]

[1]Department of Information Technologies, Murom Institute of Vladimir State University, 23 Orlovskaya Street, Murom, 602264, Russia
[2]Institute of Radio Electronics and Information Technologies, Nizhny Novgorod State Technical University n.a. R.E. Alekseev, 24 Minin Street, Nizhny Novgorod, 603950, Russia
[3]Division of Electronics and Electrical Engineering, Dongguk University, 30 Pildong-ro 1-gil Jung-gu, Seoul, 04620, Republic of Korea
[4]Laboratory of Stochastic Multistable Systems, Lobachevsky State University of Nizhny Novgorod, 23/3 Gagarin Avenue, Nizhny Novgorod, 603022, Russia

E-mail: bordanov2011@yandex.ru

Abstract. Here we present an analysis of the current state in the field of development of hardware accelerators of artificial intelligence (AI). Despite the fairly good progress made over the past decades, this area is experiencing a number of significant difficulties in its development. The solution to this problem lies in the application of new approaches to the organization of computing, in particular, computing in memory enabled by nanoelectronic memristive devices. We provide an overview of state-of-art systems, as well as our own version of the experimental concept of AI accelerators based on metal-oxide memristive devices and the massively parallel architecture for information processing.

1. Introduction

Artificial intelligence (AI) accelerator is a class of specialized hardware aimed at running machine learning algorithms and models (especially artificial neural network (ANN) models) as faster as it possible. With the development of AI systems, humanity is acquiring the ability to solve problems that are hard to formalize and perhaps even conceptualize. The fields of application of AI accelerators are robotics, uncrewed vehicles, medical equipment, web and media applications, scientific research, etc. Many countries have their own AI development programs, as this issue is strategic and is included in the list of technologies that ensure the technological independence and competitiveness of countries in the rapidly changing world.

The models and algorithms underlying the AI systems are quite complex themselves and require large computational costs. To perform these algorithms, the best achievements from the field of microelectronics and system engineering are currently being applied. Despite this, a global challenge on the way to the development of AI accelerators is the strong interdependence between the performance of electronic hardware, its size and energy consumption. This is especially critical for onboard, embedded or wearable devices, for which it is important to have small dimensions and low power consumption. In such cases, devices have to work in thin client mode, collect data, and process it on remote computers using cloud technologies.

The achieved level of performance of modern processors is sufficient to solve a huge number of tasks in which mass parallelism of information processing is not required, unlike the AI systems. Therefore, at present, the main efforts in this area should be focused on the optimization of computing hardware to improve the performance of specific AI models and algorithms. In this paper, we briefly describe the current state in the field of development of AI accelerators. In Section II, we consider the main advances in AI acceleration, highlighting the challenges and walls on the way to its development. In Section III, one can find information about the role of memristive devices in AI acceleration and about how they can improve the characteristics of accelerators. In Section IV, we specify some details on the experimental concept of memristive AI accelerator based on massively parallel architecture before concluding the paper (section V).

2 AI accelerators: achievements and challenges

The main mathematical operations of modern AI models based on ANN architectures are the matrix-vector multiplication and convolution performed at the neurons. The advantages of the ANN architecture allow processing data in parallel at all neurons in layers, while in one layer there can be hundreds of neurons at which millions of multiplications and additions are performed. Therefore, the performance of AI accelerators depends on the size and bit depth of vectors and matrices, which can be multiplied in one clock signal.

An arithmetic logic unit of a processor performs multiplication and additions operations step-by-step, and this process lasts much longer than it could last with parallel operating neurons. An additional limitation is that the processor has to access the memory for ANN weights every time it has to multiply the input signal in a neuron. This leads to the fact that the arithmetic logic unit spends most of the time not on the matrix-vector multiplication and convolution operations, but on preparing for them, thereby increasing not only the time spent, but also the energy spent [1].

The main characteristics of processors have been improved for many years due to step-by-step reduction in the size of transistors fabricated using CMOS (complementary metal-oxide-semiconductor) technology. However, currently many scientists note that Moore's law, Denard's scaling, Koomey's law and other trends in microelectronics have ended [2]. Therefore, at present, the main ways to accelerate AI models running are to create universal processors with multiple computing cores performing arithmetic and logical operations in parallel or to create application-specific integrated circuits (ASICs).

Over the last few years, many detailed surveys and benchmarks of AI accelerators have been published [3–6]. They consider more than 50 models of AI accelerators, starting from very low power chips for embedded systems to data center systems consisting of dozens of processors. Main characteristics to be compared: peak performance, peak power consumption, instructions per second, utilization percent, etc. In accordance with these surveys, graphic and tensor processors units (GPU and TPU) and multiprocessor systems based on them currently have clear advantages in all characteristics.

As the analysis of these articles has shown, the parallel calculation of multiplication and addition operations can significantly speed up the launch of AI models and improve the characteristics listed above. However, fundamental problems related to the digital approach to performing arithmetic operations on transistor logic and the von Neumann architecture remain unsolved [7]. To solve these problems, new materials, devices and computing architectures are being actively explored at the moment, which are designed to complement and possibly replace conventional devices and circuits based on CMOS technology.

3. Memristive devices: prospects in AI

Memristors are devices that have the ability to reversibly change their conductance under the action of an electric charge flowing through them [8]. This behavior caused by the resistive switching effect [9], inherent to a number of organic [10,11] and inorganic [12,13] materials and structures based on them, from which memristive devices are made. Currently, memristors with a metal-oxide-metal structure are of the greatest interest, since they are most compatible with CMOS technology and can be integrated with other electronic components on a single chip in a single technological process. Resistive switching in such devices occurs due to the restructuring of the atomic structure and the formation of filaments [14] in thin insulator or semiconductor films of nanometer thickness under the action of an electric current.

In AI accelerators, memristive devices perform the core calculations involving matrix and vector operations, such as matrix-vector multiplication and convolution. These operations in are performed in memristive arrays in analog form and based on natural physical laws: Ohm's and Kirchhoff's. In this case, the vector is the input data encoded in the parameters of input signals (such as amplitude or frequency), and the matrix is the weight coefficients encoded in the values of conductance of memristive devices. Such approach makes it possible to significantly increase the speed of calculations and reduce energy consumption. Moreover, it fundamentally solves the problem known as the von Neumann bottleneck, eliminating the need for ongoing data movement between the processor unit and memory, because an array of memristive devices stores the weights of ANN and process input signals at once [15].

Currently, several qualitative reviews have been published, which analyze in detail the characteristics (computing density, energy efficiency, computing accuracy, and on-chip learning capability) of ANN prototypes based on cross-bars of memristive devices of different sizes [16–19]. For instance, in the article [20], on the example of a two-layer perceptron network that solves the problem of handwritten digit recognition, it is experimentally shown that energy efficiency 19.7 and 6.5 times better than the Google's TPU and ASIC system, respectively. Another example has been published it the article [19] where the authors propose the memristor-based convolutional neural network which has an energy efficiency more than 2 orders of magnitude greater than that of state-of-the-art GPUs. The performance benchmark of this neuromorphic computing system shows 110 times better energy efficiency and 30 times better performance density compared with Tesla V100 GPU.

These studies confirm the prospects of this research and development area and show that even prototypes of memristive AI accelerators are able to compete in performance with existing neuroprocessors based on traditional digital elements and specialized architectures and are 2-3 orders of magnitude ahead of them in energy efficiency. In accordance with the roadmap [17], the creation of general-purpose memristive neuroprocessors is expected over the next 5-10 years, but there are still a large number of unsolved problems and challenges along this way related to co-optimization of memristive materials, devices and systems.

4. Memristive AI accelerator

For creating AI accelerators, we use memristive devices based on a thin-film metal-dielectric-metal structure, the design and fabrication technology of which are fully compatible with the modern CMOS process. The dielectric in this case is yttrium-stabilized zirconium dioxide. It is a transition metal oxide with a predominant ionic character of chemical bonding (known as an ion electrolyte), in which doping with yttrium oxide (8-12 mol.%) stabilizes the cubic phase of metal oxide and sets a certain concentration of oxygen vacancies, which play an important role in resistive switching. The design with different electrodes makes it possible to implement bipolar resistive switching of the filamentary type with a large ratio of resistances in extreme states and a large number of distinguishable programmable resistive states (at least 8), and smooth switching (plasticity) under the action of spike-like activity in a relatively smaller resistance range due to the implementation of a combined filament/interface switching mechanism.

The main technological method for the deposition of thin films of dielectrics and metals is the magnetron sputtering method. In order to optimize the parameters of resistive switching (the number of resistive switching cycles, the ratio of resistances in high-resistance and low-resistance states, the values of electroforming and switching voltages), a variation of the composition and geometry (layer thicknesses) in thin-film structures is provided. Combinations of materials for electrodes are selected from a number of metals with different vacancy formation energies relative to the dielectric materials used (TiN, Ti, Au, Zr, etc.), taking into account the results of theoretical calculation of the characteristics of interface processes. Thin-film memristive structures are characterized using high-resolution microscopic and spectroscopic methods and methods for studying electrophysical properties.

To carry out the investigations with cross bars of memristive devices, it is necessary to apply signals to all inputs in parallel and do the same for registration the results of matrix-vector multiplication or convolution. For these purposes we use a hardware-in-the-loop simulation system based on FPGA (Field-programmable gate array). To control the modes of operation of the FPGA, we use the MemriSim software, consisting of drivers that provide low-level interaction with the FPGA, and user software with a graphical interface. In addition to managing operating modes, the tasks of MemriSim include processing measurement results and evaluating cross-bar performance metrics [21], such as accuracy [22,23], fault tolerance [24,25] and reliability. The practical application of this system is presented in [26].

For the development of the experimental AI-accelerator architecture, the SMP (Symmetric Multiprocessing) architecture has been chosen as the basis. In this architecture, all processors are connected through a common system bus with memory and I/O (input/output) devices. This allows individual processors to work independently of each other and exchange data through the shared memory. However, taking into account the peculiarities of memristive devices and cross-bars, the SMP architecture will be significantly modernized. Instead of separate processors, separate cells of the AI accelerator (nodes) will be connected to the shared memory (Figure 1). An FPGA serving a set of memristive cross-bars will be placed on each cell. Buffers will be allocated for each FPGA in the shared memory, and the task of the FPGA will be to continuously monitor the state of the buffer and execute certain commands that set up the operating modes of the node.

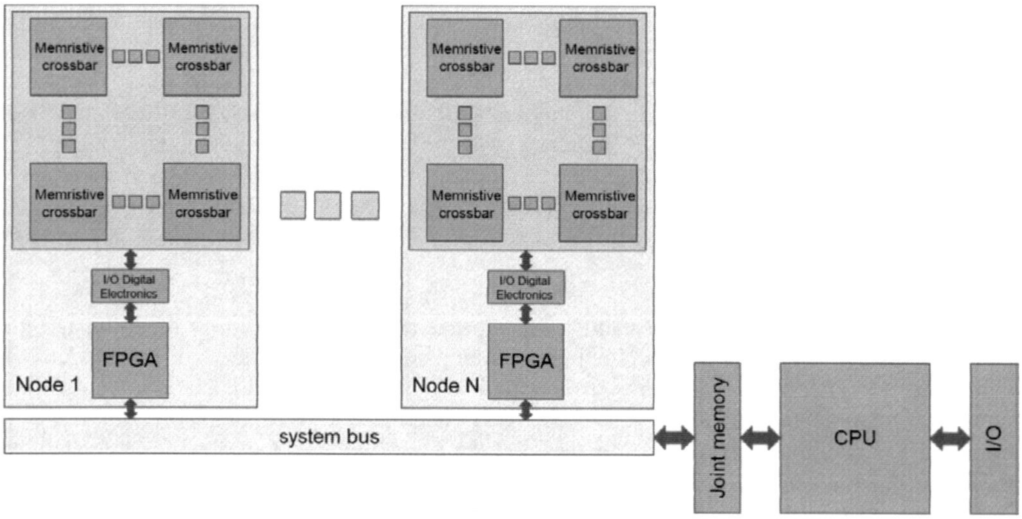

Figure 1. Scalability of the experiential AI accelerator inside a node and for the whole system.

The management of individual nodes will be implemented by the central processor, whose task is to work with the buffers of individual FPGAs. The processor will receive data from I/O devices (for example, frames of streaming video), place it in the specified buffers. At this time, the FPGA will pass this data through the cross-bars and place the result in the buffer. The result, depending on the configuration of the AI accelerator, can be either final (one node performs the functionality of one ANN) or intermediate (one node is one or more layers of the ANN). This approach opens up wide opportunities for scaling the AI accelerator both at the level of individual nodes and at the level of the entire system and makes it possible to conduct investigations of characteristics of different configurations of the whole system.

5. Conclusions

In this report, we have presented an analysis of the current state in the field of development of AI accelerators. The most popular direction of development is the use of traditional computing systems built on the von Neumann architecture, but optimized for AI tasks by increasing the number of computing cores and reducing the list of instructions they execute to the necessary and sufficient number (matrix-vector multiplication, convolution, activation, etc.). However, with such approach, these instructions are still executed digitally, and the weights of neurons and convolution cores of ANNs are still stored in a separate memory area.

We see the prospect of AI hardware accelerators developed in a different way – by applying neuromorphic and bio-inspired principles. These prospects for have been opened thanks to the discovery of memristive devices. This task requires an interdisciplinary solution step by step and is presented in the roadmap [27]. Here the experimental concept of AI accelerator has been proposed based on existing and future solutions in the field of multiprocessor computer hardware and memristive devices for real-time data processing and this concept is one of the steps toward the creation of compact, fast and energy efficient AI accelerators that cannot be realized on the basis of traditional architectures and devices.

Acknowledgments

This work was supported by the Ministry of Science and Higher Education [Agreement No. 075-15-2021-1017].

References

[1] Horowitz M 2014 *Digest of Technical Papers - IEEE Inter. Solid-State Circuits Conference* 14147629
[2] Theis T N and Wong H-S P 2017 *Comput Sci Eng* **19** 41–50
[3] Reuther A, Michaleas P, Jones M, Gadepally V, Samsi S and Kepner J 2021 AI Accelerator Survey and Trends *Preprint* 2109.08957v1
[4] Reuther A, Michaleas P, Jones M, Gadepally V, Samsi S and Kepner J 2020 *IEEE High Performance Extreme Computing Conf.* 20299956
[5] Li W and Liewig M 2020 A Survey of AI Accelerators for Edge Environment 35–44
[6] Wang Y, Wang Q, Shi S, He X, Tang Z, Zhao K and Chu X 2020 *20th IEEE/ACM International Symposium on Cluster, Cloud and Internet Computing* 744–51
[7] Zidan M A, Strachan J P and Lu W D 2018 *Nat. Electron.* **1** 22–9
[8] Chua L O 1971 *IEEE Trans. Circuits Syst. I: Regul. Pap.* **18** 507–19
[9] Ielmini D and Waser R 2016 *Resistive Switching: From Fundamentals of Nanoionic Redox Processes to Memristive Device Applications* (Weinheim: Wiley-VCH) p 784
[10] Battistoni S 2021 *BioNanoScience* **11** 1–5
[11] Gerasimov Y, Zykov E, Prudnikov N, Talanov M, Toschev A and Erokhin V 2021 *Chaos Solitons Fractals* **143** 110549
[12] Mikhaylov A, Belov A, Guseinov D, Korolev D, Antonov I, Efimovykh D, Tikhov S, Kasatkin A, Gorshkov O N, Tetelbaum D, Bobrov A, Malekhonova N, Pavlov D, Gryaznov E and Yatmanov A 2015 *Mater. Sci. Eng. B:* **194** 48–54
[13] Manouras V, Stathopoulos S, Garlapati S, Serb A and Prodromakis T 2021 *IEEE Trans. Electron Devices*
[14] Guseinov D, Tetelbaum D, Mikhaylov A, Belov A, Shenina M, Korolev D, Antonov I, Kasatkin A, Gorshkov O N, Okulich E, Okulich V, Bobrov A, Malekhonova N, Pavlov D, Gryaznov E and Gryaznov E 2017 *International Journal of Nanotechnology* **14** 604–17

[15] Xia Q and Yang J J 2019 *Nat. Mater.* **18** 309–23
[16] Amirsoleimani A, Alibart F, Yon V, Xu J, Pazhouhandeh M R, Ecoffey S, Beilliard Y, Genov R and Drouin D 2020 *Adv. Intell. Syst.* **2** 2000115
[17] Zhang W, Gao B, Tang J, Yao P, Yu S, Chang M F, Yoo H J, Qian H and Wu H 2020 *Nat. Electron.* **3** 371–82
[18] Erokhin V 2020 *BioNanoScience* **10** 834–47
[19] Yao P, Wu H, Gao B, Tang J, Zhang Q, Zhang W, Yang J J and Qian H 2020 *Nature* **577** 641–46
[20] Sun Z, Pedretti G, Bricalli A and Ielmini D 2020 *Sci. Adv.* **6** eaay2378
[21] Danilin S N, Makarov V V and Shchanikov S A 2014 *IEEE Inter. Conf. on Engineering and Telecommunication* 67–71
[22] Danilin S N and Shchanikov S A 2015 *IEEE Inter. Conf. on Mechanical Engineering, Automation and Control Systems* 15805248
[23] Danilin S N, Makarov M V and Shchanikov S A 2014 IEEE *Inter. Conf. on Mechanical Engineering, Automation and Control Systems* 14823119
[24] Danilin S N, Shchanikov S A, Sakulin A E and Bordanov I A 2018 *Proc.– 5th Inter. Conf. on Engineering and Telecommunication,* 205–9
[25] Danilin S N, Makarov M V and Shchanikov S A 2015 *Appl Mech Mater* **756** 507–12
[26] Shchanikov S, Bordanov I, Zuev A, Danilin S, Korolev D, Belov A and Mikhaylov A 2021 *BioNanoScience* **11** 84–90
[27] Mikhaylov A, Pimashkin A, Pigareva Y, Gerasimova S, Gryaznov E, Shchanikov S, Zuev A, Talanov M, Lavrov I, Demin V, Erokhin V, Lobov S, Mukhina I, Kazantsev V, Wu H and Spagnolo B 2020 *Front Neurosci* **14** 358

Journal of Physics: Conference Series **2316** (2022) 012002 doi:10.1088/1742-6596/2316/1/012002

Zero-field level-crossing resonances in a small low-temperature cesium vapor cell for atomic magnetometry applications

D V Brazhnikov[1,2], V I Vishnyakov[1], C Andreeva[3,4] and A N Goncharov[1,2,5]

[1] Institute of Laser Physics SB RAS, 15B Lavrentyev Ave., 630090 Novosibirsk, Russia
[2] Novosibirsk State University, 1 Pirogov Str., 630090 Novosibirsk, Russia
[3] Institute of Electronics BAS, 72 Tzarigradsko Chaussee Blvd., 1784 Sofia, Bulgaria
[4] Faculty of Physics, Sofia University "St. Kliment Ohridski", 1504 Sofia, Bulgaria
[5] Novosibirsk State Technical University, 20 Karl Marks Ave.,630073 Novosibirsk, Russia

E-mail: brazhnikov@laser.nsc.ru

Abstract. Zero-field level-crossing resonances (LCRs) underlay one of the most robust and simplest techniques in atomic magnetometry. LCR-based miniaturized magnetic field sensors have already found relevant applications in biology and medicine. Such sensors utilize a single circularly polarized light wave to interact with alkali-metal atoms (usually Rb or Cs) and to observe the LCR in a vapor-cell transmission when a transverse magnetic field is scanned around zero value. A high-temperature regime ($T \approx 120 - 160^\circ$C) is required to achieve a desirable sensitivity of measurements. It can be a problem for some applications, especially in a multi-channel mode of operation. Here, we consider two spectroscopy techniques that can provide high-contrast and narrow LCRs under a relatively low temperature of the cell ($\leq 60^\circ$C). These techniques imply using two light waves: the pump wave to polarize the atoms and the probe one to register the resonance. A cubic glass cell of $5\times5\times5$ mm^3 size is used in the experiments. It is filled with cesium vapors and neon as a buffer gas. The results can be used for developing a miniaturized low-power high-sensitivity magnetic field sensor for biomedical applications.

1. Introduction

The Hanle effect [1] and the underlying zero-field level-crossing resonances [2] have played a significant role in the development of quantum mechanics. They have found many interesting applications in plasma diagnostics, physics of collisions between atoms and electrons, measuring excited-state lifetimes of atoms, astrophysics and so on (see reviews [3, 4]). Perhaps, one of the most relevant applications lays in the field of quantum magnetometry. Such magnetic field sensors are based on the LCR in an atomic ground state – the so-called ground-state Hanle effect (GSHE). By using a buffer gas or an antirelaxation coating of the cell walls, GSHE resonance's linewidth can be extremely small, providing high sensitivity of measurements. The possibility of this was first proposed and studied back in the late 1960s [19, 20].

The simplicity of GSHE-based atomic spectroscopy technique has led to the development of miniature atomic magnetometers (AMs), utilizing tiny alkali-metal vapor cells [7]. These magnetometers are nowadays successfully applied to magnetoencephalography (MEG) [8, 9], magnetocardiography (MCG) [10, 11] and biomagnetism [12] studies. To achieve the best

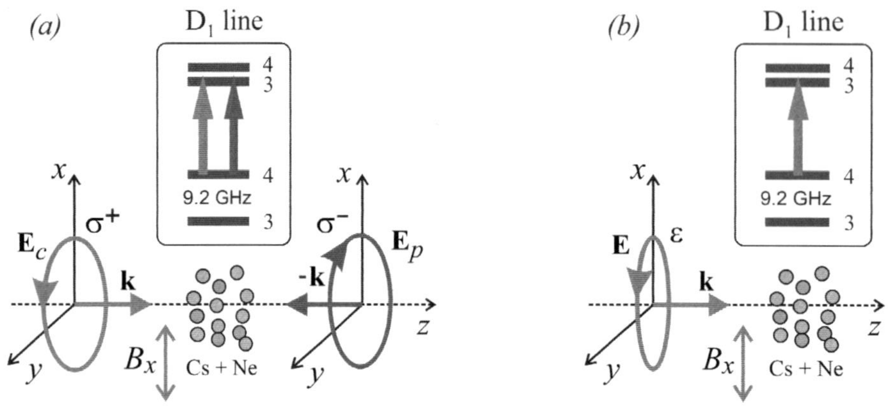

Figure 1. (a) The light-field configuration composed of two counterpropagating orthogonally circularly polarized beams. $\mathbf{E}_c$, $\mathbf{E}_p$ stand for the pump and for the probe fields, respectively. (b) The configuration with a single elliptically polarized light wave. In both cases, the transverse magnetic field B_x is scanned to observe the zero-field level-crossing resonance. The insets present the cesium D_1 line where numbers mean the total angular momenta F in the excited and ground states.

performance, the GSHE-based sensors require heating vapor cells to relatively high temperatures up to $\approx 160°$C. Under this condition, a so-called spin-exchange-relaxation-free (SERF) regime is activated (e.g., see review [13]). However, this regime requires high suppression of an ambient magnetic field and is mainly used for near-zero-field measurements (usually less than 100 nT). Heating to high temperatures also means relatively large heat release near the cell and high electric power consumption of the sensor as a whole that can bring some limitations, especially in a multi-channel regime of operation when an array of sensors is engaged.

A standard Hanle configuration that underlies state-of-the-art miniature GSHE-based sensors is quite simple and utilizes a single circularly polarized light wave. A transverse magnetic field ($\mathbf{B}_x \perp \mathbf{k}$) is scanned to observe a LCR in the light wave intensity transmitted through the cell (with $\mathbf{k}$ being the wave vector). In this work, we consider two pump-probe light-field configurations that allow obtaining high-quality LCRs under a relatively low temperature of the cell ($\leq 60°$C). As with the standard Hanle scheme, the transverse magnetic field is scanned to observe the resonance in the probe wave intensity.

The first scheme uses two counterpropagating light beams with opposite circular polarizations (figure 1a). The other scheme involves only one elliptically polarized light wave that can be decomposed into two copropagating circularly polarized light waves (figure 1b). In both cases, the waves are in resonance with the cesium D_1 line's ground-state level F_g=4. The excited-state hyperfine levels F_e=3, 4 are not spectroscopically resolved due to collisions with a buffer gas that is presented in the cell (≈ 120 Torr of Ne). We use a small cubic vapor cell with $5\times5\times5$ mm^3 dimensions made of the borosilicate ("pyrex") glass . It is heated by an ac (100 kHz) electric current applied to special heating elements. As will be shown below, both schemes demonstrate high contrast of the resonances (up to $\approx 99\%$ in the counterpropagating beams), accompanied by a relatively small linewidth of the order of 1 mG (100 nT). The contrast-to-width ratio can be considered as a quality factor of the resonance that determines the sensitivity of magnetic field measurements. In our experiments, the optimal quality factor lays in the range 30$-$55 %/mG and is competitive with those obtained in other works where the standard Hanle configuration was used. However, the main advantage of the schemes proposed consists in much lower temperature of the vapor cell, so that the sensor can be placed closer to the object, yielding larger signals and sensitivity. The results, therefore, can be applied to the development of a low-power high-sensitivity miniaturized magnetic field sensor.

2. Two counterpropagating circularly polarized light waves

In the GSHE-based AMs, either linearly [14, 15, 16, 17, 18] or circularly [19, 20, 21] polarized light beam can be used for magnetic field measurements. In the former case, under zero magnetic field, the alkali atoms experiences a macroscopic vector magnetization (spin orientation) in the cell, while in the latter case a tensor magnetization (spin alignment) takes place. Then, if the magnetic field is non-zero, the light-induced atomic polarization moments experience precession, affecting the light absorption in the cell that can be measured. Although both linear and circular beam polarizations can be applied to magnetic field measurements, the most compact sensor heads are based on a configuration with the circularly polarized beam where a transverse magnetic field is scanned to observe the LCR [7, 8, 9, 10, 11, 12, 22].

Here, we propose a slight modification of the standard scheme by means of adding the second (pump) circularly polarized light wave traveling in an opposite direction to the first (probe) beam. Both beams are of the same optical frequency ω and are in resonance with the D_1 line. The similar pump-probe scheme composed of two counterpropagating linearly polarized light waves, here referred to as "$lin\perp lin$", has already demonstrated high-quality LCRs in rubidium [23] as well as in cesium [24, 25] buffered vapor cells. The observation of high-contrast electromagnetically induced absorption (EIA) resonances in the probe wave transmission was powered by the creation of a coherent-population-trapping (CPT) state [26] that was a "dark" state for the pump beam, while it was a "bright" state for the probe one [27, 28, 29]. Let us note that, in the standard scheme with a single circularly polarized light wave, the "dark" state is usually a reason for observation of an "opposite" effect - the electromagnetically induced transparency (EIT).

Unfortunately, the $lin\perp lin$ configuration could not be successfully applied to miniaturized vapor cells where a higher buffer gas pressure is required. We, therefore, proposed to use the circularly orthogonally polarized beams ("$\sigma^+\sigma^-$") instead of the linearly polarized ones [30]. This configuration is based on a spin orientation of atoms by the pump wave. The spin orientation process is not only destroyed, but it becomes even more efficient under a higher buffer gas pressure in contrast to the CPT-state creation, which is highly sensitive to overlapping the upper-state hyperfine levels due to the rapid collisions of the alkali atoms with the buffer gas atoms. Besides, the $\sigma^+\sigma^-$ scheme does not exhibit an additional shift of the resonance [31] that, in our case, can be considered as an unwanted side effect.

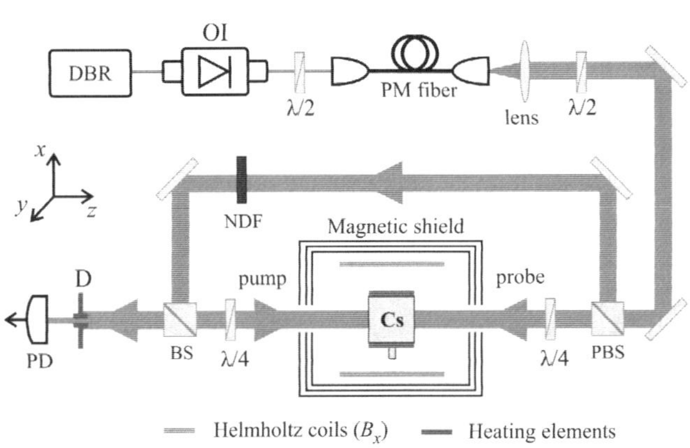

Figure 2. The experimental setup for $\sigma^+\sigma^-$ configuration: DBR, distributed Bragg reflector diode laser; OI, optical Faraday insulator; $\lambda/2$, $\lambda/4$, half-wave and quarter-wave plates, respectively; PM fiber, polarization maintaining optical fiber; PBS, polarizing beam splitter; BS, beam splitter; NDF, set of neutral density filters; D, iris diaphragm; PD, photodetector.

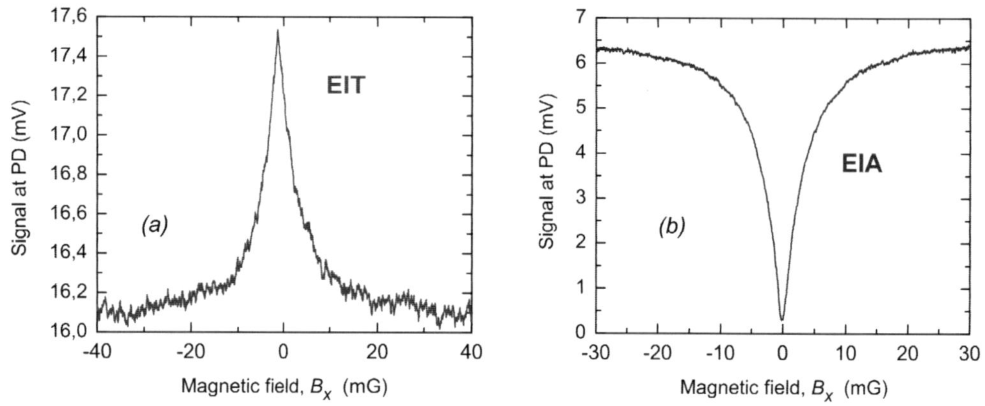

Figure 3. (a) Level-crossing resonance in the standard scheme with a single circularly polarized light wave when an EIT-like resonance is observed, $P_p \approx 1$ mW. (b) EIA-like level-crossing resonance in the proposed $\sigma^+\sigma^-$ scheme, $P_p \approx 5$ μW, $P_c \approx 1$ mW. The resonances were monitored at $T_{cell} \approx 55°$C.

Inspired by the results of [30], here we continue investigating the $\sigma^+\sigma^-$ scheme and show how a contrast-to-width ratio can be further improved by means of such a simple element as an iris diaphragm installed on the probe beam's path. The experimental setup is shown in figure 2. A cubic Cs vapor cell is placed inside a three-layer μ-metal magnetic shield that reduces the stray field to ≈ 1 mG in the cell area.

We use a distributed Bragg reflector (DBR, Toptica LD-0895-0040) diode laser with a radiation wavelength of $\lambda = 894.6$ nm and a linewidth of ≈ 0.5 MHz. The laser frequency is tuned manually to the center of the combined absorption profile, corresponding to the optical transitions $F_g=4 \rightarrow F_e=3$ and $F_g=4 \rightarrow F_e=4$. The laser beam diameter ($1/e^2$) after a lens is around 3 mm. The combination of a half-wave plate placed after the lens and a polarizing beam splitter (PBS) is used to redistribute the total light power between the pump and probe beams, which have orthogonal linear polarizations after the PBS. A beam splitter (BS) is used to direct the pump beam into the cell without affecting its polarization. A set of neutral density filters (NDF) is used to tune the pump wave power. Two quarter-wave plates ($\lambda/4$) are placed on both sides of the cell to produce the desirable $\sigma^+\sigma^-$ configuration. A pair of Helmholtz coils is used to produce a transverse magnetic field (B_x), which is scanned around zero value to observe the LCR in the probe wave intensity. In contrast to our previous work [30], here we use an iris diaphragm (D) on the probe wave path. Argumentation in favor of using this tool has been provided in [24], therefore, we do not reproduce it in the current paper.

Figure 3 demonstrates examples of LCRs in the standard and $\sigma^+\sigma^-$ configurations under a low-temperature regime. As seen, the EIA-like resonance has much higher contrast then the EIT one. We define contrast as a ratio between the resonance's height and a value of background transmission at the resonance's wings. Figure 4a reveals that the EIA resonance's contrast can reach 99% very fast with the pump wave intensity increase, while in the EIT case the contrast is no more than $\approx 10\%$. For applications in atomic magnetometry, a contrast-to-width (CTW) ratio plays an important role for increasing the sensitivity of measurements. Therefore, we show this ratio in figure 4b where an optimal pump (probe) wave intensity can be figured out. It is also seen that the CTW ratio is negligibly small in the standard scheme compared to that in the proposed configuration.

Finally, figure 5 proves a positive role of the iris diaphragm. In particular, it can bring a

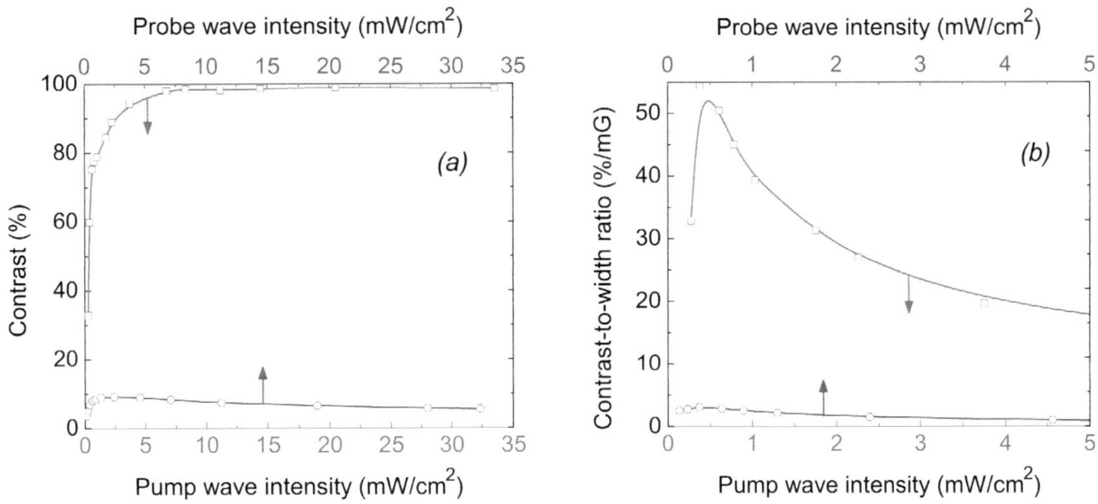

Figure 4. Contrast (a) and Contrast-to-width ratio (b) of the LCR in the standard (green circles) and $\sigma^+\sigma^-$ (pink squares) schemes at $T_{cell} \approx 55°$C.

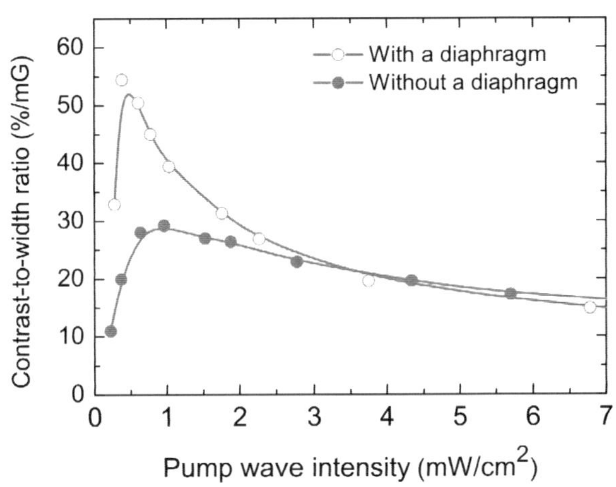

Figure 5. Influence of an iris diaphragm on the contrast-to-width ratio of EIA resonance in the $\sigma^+\sigma^-$ scheme. $T_{cell} \approx 55°$C.

two-fold enhancement in the CTW ratio.

3. A single elliptically polarized light wave

Linear and circular polarizations are the most typical light wave polarizations that are used in various atomic magnetometry methods. However, there are several proposals where an elliptically polarized light wave have successfully been implemented into an atomic magnetometer [32, 33, 34, 35]. Here, we propose another one that can be treated as a single-beam analogue of the $\sigma^+\sigma^-$ scheme described in section 2. Obviously, such a scheme is easier to implement than the $\sigma^+\sigma^-$ scheme (see figure 1b). If the x-axis is directed along the major axis of the light polarization ellipse, then the elliptically polarized wave of frequency ω can be decomposed into two circularly polarized waves as follows:

$$\mathbf{E}(z,t) = E_{-1}\mathbf{e}_{-1}e^{i(kz-\omega t)} + E_{+1}\mathbf{e}_{+1}e^{i(kz-\omega t)}, \tag{1}$$

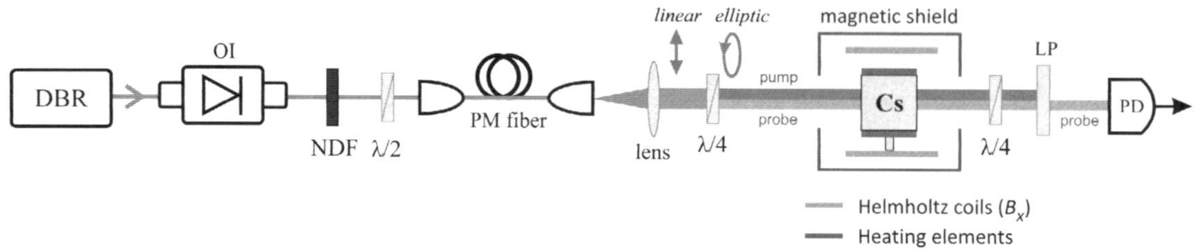

Figure 6. The experimental setup for the scheme with a single elliptically polarized beam: DBR, distributed Bragg reflector diode laser; OI, optical Faraday insulator; $\lambda/2$, $\lambda/4$, half-wave and quarter-wave plates, respectively; PM fiber, polarization maintaining optical fiber; LP, linear polarizer; NDF, set of neutral density filters; PD, photodetector.

where $k{=}\omega/c$ is the wave number with c being the speed of light, $\mathbf{e}_{\pm1}$ are spherical basis vectors responsible for the σ^+ and σ^- transitions in the atoms, respectively. The real amplitudes $E_{\pm1}$ reads:

$$E_{-1} = E_0 \sin(\varepsilon - \pi/4)\,, \qquad E_{+1} = E_0 \cos(\varepsilon - \pi/4)\,, \tag{2}$$

with ε being the ellipticity parameter, while E_0 defines the total light wave intensity as $I = cE_0^2/2\pi$. We assume that the probe wave intensity (I_{-1}) is much less than the pump wave intensity (I_{+1}), i.e. $I_{-1}{\ll}I_{+1}$. The latter means that the ellipticity ε is much closer to 45° than to −45°.

The experimental setup is shown in figure 6. It has many in common with that shown in figure 2. The first $\lambda/4$ waveplate makes the linearly polarized light to be slightly elliptic, while the second waveplate transforms the polarization back into the linear one that can be decomposed into a sum of mutually orthogonal linear polarizations of the probe and the pump waves. Then, the pump wave is blocked by a linear polarizer (LP), while the probe wave intensity is analyzed by a photodetector (PD). An example of the LCR is demonstrated in figure 7. As seen, it can exhibit a high contrast value as it has been revealed in the $\sigma^+\sigma^-$ scheme. The resonance shift can be explained by a residual magnetic field inside the magnetic shield (≈ 1 mG). FWHM of the resonance equals to ≈ 2.2 mG (220 nT). Both the power broadening and the residual magnetic field determine the resonance's linewidth (see also [30]). Better shielding thus may lead to a narrower resonance.

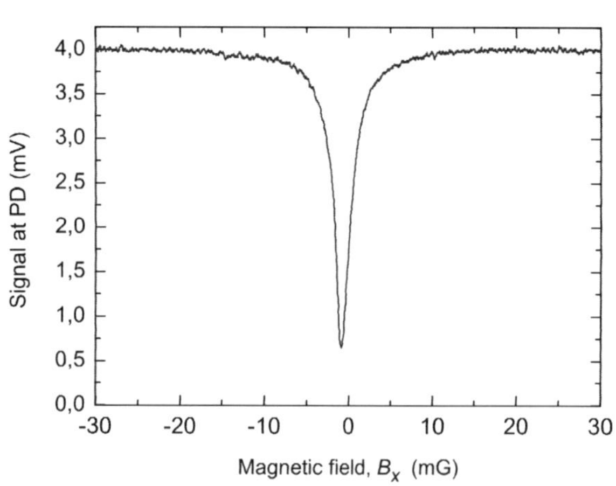

Figure 7. Level-crossing EIA resonance in the elliptically-polarized-beam scheme. The total light power is around 200 μW, $T_{cell} \approx 60$°C, $\varepsilon = 37$°.

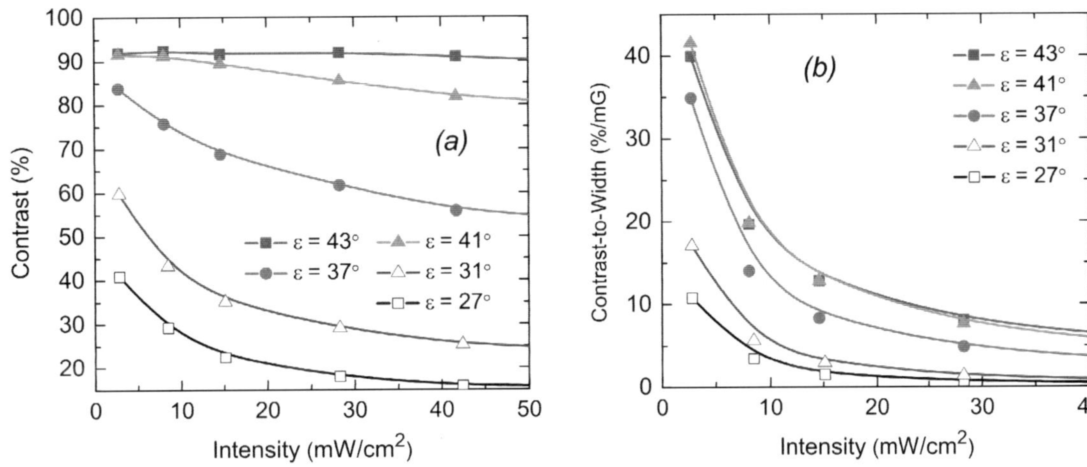

Figure 8. Contrast (a) and Contrast-to-width ratio (b) of the LCR under the elliptically polarized light with different ellipticities ε. $T_{cell} \approx 60°\text{C}$.

It should be noted that there is a distinctive feature that separates the two light field configurations depicted in figure 1 and considered here. Indeed, the counterpropagating σ^+ and σ^- waves do not lead to creation of the CPT state in the atomic ground-state level $F_g = 4$ because this state varies in space very fast as $e^{\pm 2ikz}$ and should be averaged. In the case of a single elliptically polarized light, the CPT state does not experience such fast oscillations and, at first sight, should be taken into account. However, under frequent collisions between the buffer-gas atoms and the alkali atoms, the upper-state levels $F_e = 3$, 4 are mixed and cannot be spectroscopically resolved, meaning the simultaneous excitation of both the optical transitions $F_g = 4 \rightarrow F_e = 3$ and $F_g = 4 \rightarrow F_e = 4$. These transitions are of different types according to a classification considered theoretically in [36, 37]. It means that the CPT states for these transitions strongly differs from each other and cannot survive simultaneously. In other words, mixing the upper-state levels leads to the absence of any CPT state in the ground-state level $F_g = 4$, and it does not affect the observation of nonlinear resonances. Otherwise, creation of such a state would lead to observation of the EIT-type resonance in the probe wave transmission instead of the EIA one.

Figure 8 shows the resonance contrast and CTW ratio as the functions of optical power at different ellipticities ε. The contrast as high as 92% is observed. The CTW ratio is close to that observed in the $\sigma^+\sigma^-$ scheme. The plots demonstrate that an optimal ellipticity is around 40°. The behaviour of contrast in figure 8a differs from that in figure 4a. The fact is that in the $\sigma^+\sigma^-$ scheme the probe wave power was fixed during the experiments, while in figure 8a both the pump and the probe wave intensities are changed.

4. Conclusions

We have considered two light field configurations that allow obtaining high-quality zero-field level-crossing resonances in small lower-temperature ($\leq 60°\text{C}$) buffered alkali-vapor cells. The results can be used for development of a low-power magnetic field sensor with reduced heat dissipation. The latter is especially important for applications in medicine (magnetocardiography and magnetoencephalography). The observed contrast-to-width ratio of the resonances is competitive with those obtained in other works where the SERF magnetometers were studied (e.g., see [7]). We, therefore, anticipate that the proposed techniques can achieve a sensitivity of measurements of the order of 100 $fT/Hz^{1/2}$ or even better.

Both schemes considered here have demonstrated a high quality of level-crossing resonances in the small 5 mm long cesium vapor cell. In particular, the achieved contrast-to-width ratio lays in the range 40−55 %/mG. These values are one to three orders in magnitude higher than those demonstrated in many other configurations (e.g., see [39, 40, 41, 42]). The scheme with counterpropagating σ^+ and σ^- light waves can be miniaturized similarly, for instance, to an atomic spectrometer scheme proposed in [38]. However, it is obvious that the second configuration with a single elliptically polarized light wave (figure 6) is easier to miniaturize.

Acknowledgments

The work has been supported by Russian Foundation for Basic Research (20-52-18004) and Bulgarian National Science Fund (KP-06-Russia/11) in the framework of a joint research project. We also thank Russian Science Foundation (17-72-20089).

References

[1] Hanle W 1924 *Z. Phys.* **30** 93
[2] Breit G 1933 *Rev. Mod. Phys.* **5** 91
[3] Novikov L N, Skrotskii G V and Solomakho G I 1975 *Sov. Phys. Usp.* **17** 542
[4] Gawlik W, Gawlik D and Walther H 1991 *The Hanle Effect and Atomic Physics* in *The Hanle Effect and Level-Crossing Spectroscopy*, ed G Moruzzi and F Strumia (New York: Springer).
[5] Aleksandrov E B, Bonch-Bruevich A M and Khodovoi V A 1967 *Opt. Spectrosc.* **23** 151
[6] Dupont-Roc J, Haroche S and Cohen-Tannoudji C 1969 *Phys. Lett.* A **28** 638
[7] Shah V, Knappe S, Schwindt P D D and Kitching J 2007 *Nat. Photonics* **1** 649
[8] Boto E *et al.* 2017 *NeuroImage* **149** 404
[9] Pratt E J *et al.* 2021 *Proc. SPIE* **11700** 1170032
[10] Alem O, Sander T H, Mhaskar R, LeBlanc J, Eswaran H, Steinhoff U, Okada Y, Kitching J, Trahms L and Knappe S 2015 *Phys. Med. Biol.* **60** 4797
[11] Yang Y, Xu M, Liang A, Yin Y, Ma X, Gao Y and Ning X 2021 *Sci. Rep.* **11** 5564
[12] Fabricant A, Iwata G Z, Scherzer S, Bougas L, Rolfs K, Jodko-Władzińska A, Voigt J, Hedrich R and Budker D 2021 *Sci. Rep.* **11** 1438
[13] Savukov I and Seltzer S J 2013 *Spin-Exchange-Relaxation-Free (SERF) Magnetometers* in *Optical Magnetometry*, ed Budker D and Jackson Kimball D F (Cambridge: Cambridge University Press)
[14] Alipieva E A, Gateva S V, Taskova E T 2004 *Proc. SPIE* **5449** 336
[15] Gateva S, Alipieva E, Petrov L, Taskova E and Todorov G 2008 *J. Optoelectron. Adv. Mater.* **10** 98
[16] Grewal R S and Pattabiraman M 2016 *Eur. Phys. J.* D **70** 219
[17] Zhang H B, Yang G, Huang G M and Li G X 2019 *Phys. Rev.* A **99** 033803
[18] Le Gal G, Lieb G, Beato F, Jager T, Gilles H and Palacios-Laloy A 2019 *Phys. Rev. Appl.* **12** 064010
[19] Aleksandrov E B, Bonch-Bruevich A M and Khodovoi V A 1967 *Opt. Spectrosc.* **23** 151
[20] Dupont-Roc J, Haroche S and Cohen-Tannoudji C 1969 *Phys. Lett.* A **28** 638
[21] Papoyan A, Shmavonyan S, Khanbekyan A, Khanbekyan K, Marinelli C and Mariotti E 2016 *Appl. Opt.* **55** 892
[22] Shah V K and Wakai R T 2013 *Phys. Med. Biol.* **58** 8153
[23] Brazhnikov D V, Ignatovich S M, Vishnyakov V I, Skvortsov M N, Andreeva C, Entin V M and Ryabtsev I I 2018 *Laser Phys. Lett.* **15** 025701
[24] Brazhnikov D V, Ignatovich S M, Novokreshchenov A S and Skvortsov M N 2019 *J. Phys. B: At. Mol. Opt. Phys.* **52** 215002
[25] Brazhnikov D V *et al.* 2021 *J. Phys. Conf. Ser.* **1859** 012019
[26] Arimondo E 1996 *Prog. Opt.* **35** 257
[27] Brazhnikov D V, Taichenachev A V, Tumaikin A M, Yudin V I, Ryabtsev I I and Entin V M 2010 *JETP Lett.* **91** 625
[28] Brazhnikov D V, Taichenachev A V and Yudin V I 2011 *Eur. Phys. J.* D **63** 315
[29] Brazhnikov D V, Taichenachev A V, Tumaikin A M and Yudin V I 2014 *Laser Phys. Lett.* **11** 125702
[30] Brazhnikov D V, Vishnyakov V I, Ignatovich S M, Mesenzova I S, Andreeva C and Goncharov A N 2021 *Appl. Phys. Lett.* **119** 024001
[31] Brazhnikov D, Ignatovich S, Mesenzova I, Novokreshchenov A, Goncharov A 2020 *Opt. Lett.* **45** 3309
[32] Shah V and Romalis M V 2009 *Phys. Rev.* A **80** 013416

[33] Li R, Quan W, Fan W, Xing L, Wang Z, Zhai Y and Fang J 2017 *Chin. Phys.* B **26** 120702
[34] Su S, Zhang G, Bi X, He X, Zheng W and Lin Q 2019 *Opt. Express* **27** 33027
[35] Le Gal G, Rouve L and Palacios-Laloy A 2021 *Appl. Phys. Lett.* **118** 254001
[36] Smirnov V S, Tumaikin A M and Yudin V I 1989 *Sov. Phys. JETP* **69** 913
[37] Lazebnyi D B, Brazhnikov D V, Taichenachev A V, Basalaev M Yu and Yudin V I 2015 *J. Exp. Theor. Phys.* **121** 934
[38] Knappe S A, Robinson H G and Hollberg L 2007 *Opt. Express* **15** 6293
[39] Gozzini S, Cartaleva S, Lucchesini A, Marinelli C, Marmugi L, Slavov D and Karaulanov T 2009 *Eur. Phys. J.* D **53** 153
[40] Kim H-J and Moon H S 2011 *Opt. Express* **19** 168
[41] Ravi H, Bhattarai M, Bharti V and Natarajan V 2017 *Europhys. Lett.* **117** 63002
[42] Gozzini S, Fioretti A, Lucchesini A, Marmugi L, Marinelli C, Tsvetkov S, Gateva S and Cartaleva S 2017 *Opt. Lett.* **42** 2930

Investigation of the dependence of the reflection coefficient on the modulation parameters and defects of the fiber Bragg gratings

K M Stankevich, O Y Butkovskiy, D A Kochuev, A S Chernikov and I I Privodnov

Vladimir State University, Gorky Str. 87, 600000 Vladimir, Russia

E-mail: kristina.stankevich@bk.ru

Abstract. In this article we studied a model of changes in the amplitude of the spectral density maxima and interference of the side lobes of backscattering radiation from Fiber Bragg gratings (FBGs). For this purpose a computer simulation of the matrix system of equations of coupled modes depending on the modulation depth and wavelength was carried out. The practical significance of the work lies in finding the laws of dependence of the level of oscillation on changes in the input parameters to obtain the greatest efficiency of optical fiber containing fiber Bragg gratings. An attempt is made to model technological defects and their influence on the character of FBG backscattering.

1. Introduction

Fiber Bragg gratings (FBGs) are currently one of the most important elements in various fiber optics devices. In particular, they are used as dispersion compensators in fiber optic communication lines, mirrors of fiber and semiconductor lasers, and sensitive elements of fiber optic sensors of physical quantities (temperature, tension, pressure) [1]. Nowadays, FBGs are among the most popular optical components, which are widely used as sensing elements in sensing devices and as spectral filters in fiber lasers [2,3]. Bragg gratings are widely used in the field of quantum electronics, including quantum computers, which confirms the necessity to study the radiation processes occurring in Bragg fiber gratings.

The simplest example of FBG is a periodic change in the refractive index in the core of a single-mode optical fiber [1] due to modification of the core material [4]. Modern quartz fibers can change their optical properties under the influence of intense laser radiation in the ultraviolet region of the spectrum. This effect is used to induce periodic (discrete) changes in the refractive index of the core along the length of the fiber, which leads to the formation of a Bragg grating inside the core [5].

One common method of solving the problem for finding the passed and backscattered radiation is the method of coupled modes [6,7]. In works [6,7] the method of coupled modes was applied for distribution of refractive index along a fiber, carried out according to the harmonic law.

The study of radiation parameters in fiber Bragg gratings is relevant because with the help of Bragg gratings it is possible to achieve reduction of the level of lateral radiation and amplification of laser radiation scattered on the grating.

This paper belongs to the field of laser technology.

The aim of the work was to determine the position of the reflection coefficient maximum, its depth and width, the depth and tuning of side lobes depending on the refractive index modulation coefficient maximum, as well as the influence of inevitable technological fluctuations.

The main purpose of using Bragg gratings and Bragg mirrors [8] is to obtain the highest possible reflection coefficient of the light generated by the laser and the need to achieve a high reflection coefficient with a minimum number of grating strokes. Another task of the work is related to distortions of induced gratings arising due to unavoidable fluctuations of laser radiation during induction of gratings in the fiber. The simplest possible model of which will be presented in the paper.

2. Description and discussion of the modulation results

The calculations were performed for a Bragg grating with a period Λ = 5.18938 μm at a fiber length of L = 5 mm. Amplitude of the harmonic change of the induced refractive index $n_{mod} = 4{,}1\cdot 10^{-5}\text{-}7\cdot 10^{-4}$, effective refractive index of the principal mode n_{eff} = 1.4447, average value of the induced refractive index $\hat{n}_{in}$ = 2.7, wavelength λ = 1.498-1.502 μm (a number of individual cases are discussed here), power fraction of the principal mode propagating through the fiber core η = 0.8. Initial data were taken from [1].

Calculation of the effective refractive index of the main mode with the FBG recorded in the core was performed using the technique described in [9]. We consider the distribution of the induced refractive index n_{mod} constant along the radius of the core for each case.

For a homogeneous [10,11] Bragg grating in z:

FBG period $\Lambda = const$, the coupling coefficient determining the lattice-induced interaction of the forward and backward counter waves $k(z,\lambda) = k = const$, amplitude of the harmonic change of the induced refractive index $n_{mod}(z) = n_{mod} = const$ (without introducing noise exposure), z – coordinate along the length of the fiber, the noise level of the induced modulation $\sigma(z,\lambda) = \sigma = const$.

For this case we apply the solution of the system of homogeneous differential equations of the first order in matrix form [7,12]:

$$S(Z,\lambda) = C_1(\lambda)\cdot\big(i\gamma(\lambda) - \sigma(\lambda)\big)exp\big(z\gamma(\lambda)\big) + C_2(\lambda)\cdot\big(-i\gamma(\lambda) - \sigma(\lambda)\big)exp\big(-z\gamma(\lambda)\big)$$

$$R(Z,\lambda) = C_1(\lambda)\cdot k(\lambda)\cdot exp\big(z\gamma(\lambda)\big) + C_2(\lambda)\cdot k(\lambda)\cdot exp\big(-z\gamma(\lambda)\big) \qquad (1)$$

где $C_1(\lambda) = \frac{1-k(\lambda)C_2(\lambda)}{k(\lambda)}$;

$$C_2(\lambda) = \frac{\frac{\sigma(\lambda)-i\gamma(\lambda)}{k(\lambda)}exp(L\gamma(\lambda))}{(\sigma(\lambda)-i\gamma(\lambda))\cdot exp(L\gamma(\lambda))-(\sigma(\lambda)+i\gamma(\lambda))\cdot exp(-L\gamma(\lambda))};$$

$$\gamma(\lambda) = \sqrt{\Big(\big(k(\lambda)\big)^2 - \big(\sigma(\lambda)\big)^2\Big)}.$$

Using the method of coupled modes, we can obtain a system of equations for slowly varying on the wavelength scale amplitudes propagating in the forward and reverse directions.

This system has the form [4,7]:

$$\frac{\partial R}{\partial z} = i\sigma(z)R(z) + ik(z)S(z)$$

$$\frac{\partial S}{\partial z} = -i\sigma(z)S(z) - ik(z)R(z) \qquad (2)$$

where $R(z,\lambda)$ и $S(z,\lambda)$ – slowly changing on the wavelength scale wave amplitudes propagating in the forward and reverse directions, respectively.

From system (2) it is not difficult to obtain the equation for the complex amplitude of the wave scattered in the opposite direction in the form of the Mathieu equation.

$$\frac{d^2S}{dz^2} + F(n_0 + n_{mod} \cdot \cos\omega t) \cdot S = 0 \tag{3}$$

$$n_{in} = \hat{n}_{in} + \tilde{n}_{mod} \cdot \sin\left(\frac{2\pi z}{\Lambda}\right) \tag{4}$$

Equation (3) corresponds to the equation of parametric generation of a reflected wave and if the parametric resonance condition is satisfied, the occurrence, amplification, reduction and stochasticity of the reflected wave amplitude in a medium with dissipation is possible [12].

The results of modeling the dependence of the reflection coefficient on the wavelength are shown in Figure 1.

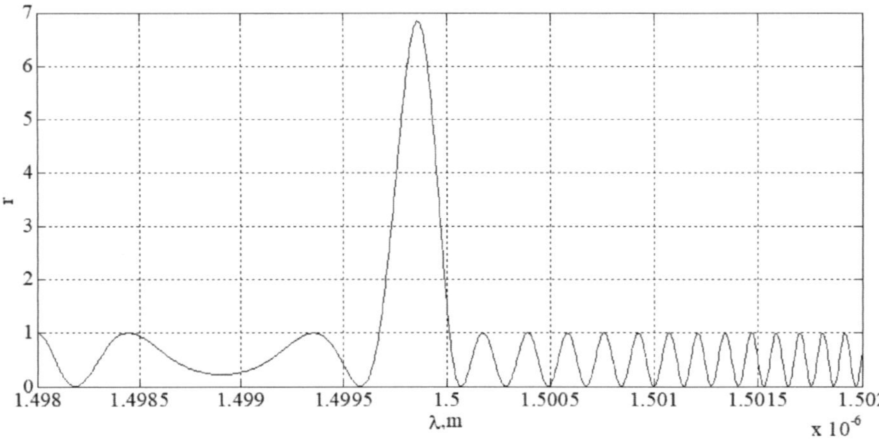

Figure 1. Dependence of the reflection coefficient of homogeneous FBG on λ.

To solve the second problem posed in this work, "technological noises" [14,15] were introduced into the refractive index modulation amplitude model under study to evaluate their effect on the wavelength-dependent reflection coefficient of the grating (Figure 2).

For this purpose, we introduced into the model the technological errors of the lattice in the form of a uniform random process, which can be associated with inevitable fluctuations of the laser intensity and processes during the formation of the lattice in the fiber. The most suitable in our opinion is the amplitude of the induced refractive index in the form [1,7,16]:

$$\tilde{n}_{mod} = n_{mod} + \sigma\xi \tag{5}$$

where $\tilde{n}_{mod}$ – amplitude of the induced refractive index with the addition of noise; σ – noise level of induced modulation; ξ – uniform noise with zero mean.

Thus, the induced refractive index in the equation for the reflection coefficient FBG [17,18]:

$$n_{in} = \hat{n}_{in} + \tilde{n}_{mod} \cdot \sin\left(\frac{2\pi z}{\Lambda}\right) \tag{6}$$

where $\hat{n}_{in}$ - average value of the induced refractive index.

The behavior of the induced refractive index as one moves along the length of the core is shown below (Figure 2). The number of strokes at a length of 48 μm is 96.

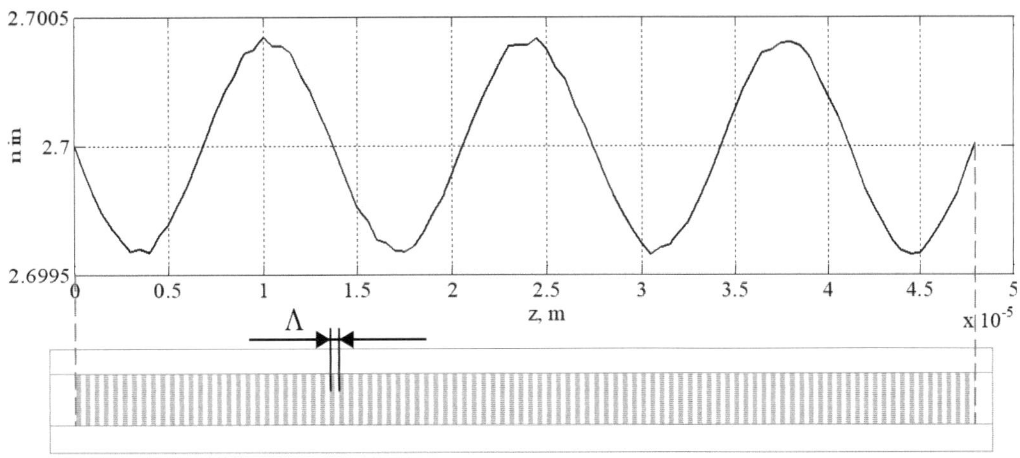

Figure 2. Diagram of changes in the induced refractive index as a function of lattice length.

The result of modeling the dependence of the reflection coefficient on the wavelength, taking into account technological fluctuations, looks like (Figure 3):

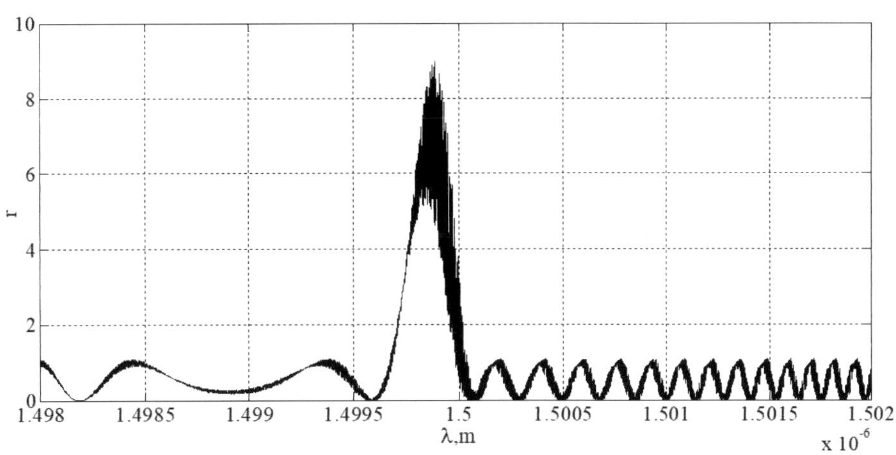

Figure 3. Reflection coefficient spectrum of FBG obtained after introducing noise into refractive index modulation.

As can be seen in Figure 3, the maximum value of the reflection coefficient increased due to random noise effects. At the same time, averaging over 15 realizations showed that the value of the average maximum of the central peak and the amplitudes of the side lobes did not change.

Next, consider the dependence of the amplitude of the side lobes on the amplitude of the harmonic change in the induced refractive index n_{mod} and the levels of maxima for each n_{mod}, given the noise effect (Figure 4).

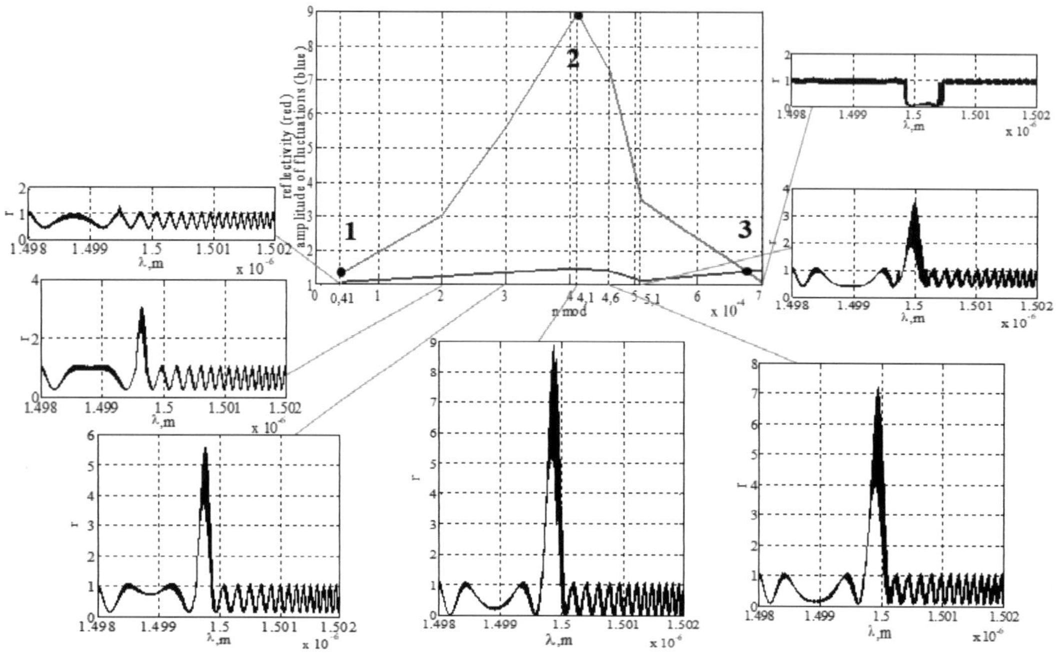

Figure 4. Diagram of variation of backscatter level (red), side lobe level (blue) with noise as a function of n_{mod}. The inserts show typical examples of simulated backscatter level dependencies for a given value of modulation amplitude.

As can be seen in Figure 4, up to point 1 on the graph, the backscatter oscillates, and there is no highlighted backscatter (as in Figure 3).

At the value of $n_{mod} = 4{,}1\cdot10^{-5}$ the main maximum becomes higher than the level of the side lobes. The values of the contrast ρ of the lattice vary from $0{,}76\cdot10^{-5}$ to $5{,}6\cdot10^{-5}$.

After point 1, the spectral density maximum becomes larger than the noise level, and the spectral power fluctuations decrease (become narrower). Increase n_{mod} increases the reflection coefficient and broadens the reflection bandwidth, but the reflection in the side maxima, caused by multiple re-reflections of the wave on non-equidistant inhomogeneities, increases insignificantly [7]. The ratio of the level of the reflection coefficient peak to the level of lateral fluctuations with an increase in the level of modulation amplitude is expected to grow. The peak maximum (point 2 in Figure 4) is observed at $n_{mod} = 4{,}1\cdot10^{-4}$ at the contrast of the grid $7{,}6\cdot10^{-5}$.

A further increase in the modulation amplitude and, consequently, an increase in the lattice contrast after point 2 leads to a decrease in the ratio of the backscattering level to the level of the side lobes. In other words, the spectral power fluctuations increase, which leads to oscillation and reduction of the spectral density level. It can be seen that the reduction of the peak is not uniform. The values of ρ vary from $8{,}5\cdot10^{-5}$ to $9{,}4\cdot10^{-5}$.

When the modulation amplitude increases above $7\cdot10^{-4}$ at $\rho = 12{,}96\cdot10^{-5}$ after point 3 in Figure 4, the backscattering disappears (the intensity of the scattered wave becomes zero).

3. Conclusion

A computer simulation of the dependence of the oscillation level on the induced refractive index was performed.

Analysis of the reflection coefficient shows that depending on the modulation depth and wavelength, the dynamics of the maximum of the power spectrum and fluctuations associated with side-blade interference are observed.

It is proved that the reflection coefficient depends not only on the Bragg wavelength and the wavelength of the input radiation, but also on the modulation associated with the modulation of the refractive index of the light guide.

The patterns of influence of FBG defects and noise level on the reflection coefficient and efficiency were revealed.

It was calculated that by increasing the lattice contrast to a certain value, the maximum level of backscattering coefficient is reached, and by further increasing the contrast, the backscattering disappears.

In general, a satisfactory model of the change in the amplitude of the spectral density maxima and the interference of the side lobes was obtained and studied.

Acknowledgments
The study was carried out using the equipment of the interregional multispecialty and interdisciplinary center for the collective usage of promising and competitive technologies in the areas of development and application in industry/mechanical engineering of domestic achievements in the field of nanotechnology (Agreement No. 075-15-2021-692 of August 5, 2021).

This work was also partially supported by the framework of the state task of VlSU № 0635-2020-0013.

References
[1] Ovvyan A and Khorokhorov A 2012 *Youth Sci-Tech. Newsletter* **77** 51038
[2] Hill K O, Fujii Y, Johnson D C and Kawasaki B S 1978 *Appl. Phys. Let.* **32** 647
[3] Schaffer C B, Brodeur A and Mazur E 2001 *Measure. Sci. Tech.* **12** 1784–94
[4] Chernikov A S, Khorkov K S, Kochuev D A, Chkalov R V, Prokoshev V G and Davydov N N 2019 *J. Phys.: Conf. Ser.* **1164** 012015
[5] Othonos A 1997 *Rev. Sci. Instrum.* **12**(68) 4309–41
[6] Erdogan T 1997 *J. of Lightwave Tech.* **8**(15) 1277–83
[7] Bugrov V, Malakhov V and Raevsky A 2010 *Wave Proc. Phys. and Rad. Engineer. Sys.* **4**(13) 25–8
[8] Masterov R and Karpov S 2021 *J. of Tech. Phys.* **12**(91) 1998–2000
[9] Raevsky S and Smirnov A 2006 *Antennas* **5**(108) 68–72
[10] Loh W and Laming R 1995 *Electron. Lett.* **17**(31) 1440–42
[11] Eggleton B, Krug P, Poladian L and Ouellette F 1994 *Electron. Lett.* **19**(30) 1620–22
[12] Sipe J, Poladian L and Sterke C 1994 *J. Opt. Soc. Amer. A.* **4**(11) 1307–20
[13] Bogoliubov N and Mitropolskiy Yu 1961 *Asymptotic Methods in the Theory of Non-Linear Oscillations (*New York: Gordon & Breach)
[14] Kurnosov V and Kurnosov K 2015 *Quantum Electron.* **1**(45) 11–4
[15] Zholnerov V, Ivanov A and Kurnosov V 2013 *Quantum Electron.* **9**(43) 824-27
[16] Yariv A 1973 *IEEE J. Quantum Electron.* **9**(9) 919–33
[17] Sterke C, Broderick N, Eggleton B and Steel M 1996 *Opt. Fiber Tech.* **3**(2) 253–68
[18] Kivshar S, Agrawal P 2003 *Optical Solitons: From Fibers to Photonic Crystals* (Academ. Press)

Synthesis of copper nanoparticles by laser ablation

A A Antipov [1, 2], A G Putilov[1, 2], A V Osipov[1,2], A E Shepelev[1], V N Glebov[1] and A M Maliutin[1]

[1]ILIT RAS – Branch of FSRC "Crystallography and Photonics" RAS,
1 Svyatoozerskaya, Shatura, 140700, Moscow region, Russia
[2] Vladimir State University, Gorky Str. 87, 600000 Vladimir, Russia

E-mail: antiplit@ya.ru

Abstract. The synthesis of copper nanoparticles using a pulsed alexandrite laser is demonstrated. The laser operated in a free generation mode with a pulse duration of about 250 μs. The copper target was placed in distilled water and glycerin. The obtained colloidal solutions were investigated by dynamic light scattering, and the deposited structures were studied by atomic force and electron microscopy. The concentration of nanoparticles depended on the power of the laser radiation, the scanning speed and the exposure time. The smallest particles size was obtained at a frequency of 5 Hz at a wavelength of 750 nm. Colloids obtained in distilled water contain copper nanoparticles, which oxidize over time.

1. Introduction

The study of the synthesis of copper nanoparticles is due to their unique properties in comparison with bulk materials and their application in various fields. Recently, copper nanoparticles have found application in the formation of conductive films [1], the creation of nonlinear devices [2] and bactericidal agents [3]. Laser ablation has an advantage over chemical synthesis methods. By this method, the synthesis takes place in a short time, chemically pure substances are formed, and hybrid materials of dissimilar metals can also be obtained. Also, the synthesized nanoparticles have high activity and are almost immediately ready for use [4, 5].

2. Methods and equipment

Copper is a plastic material of golden color with good electrical conductivity and high reflectivity in the near infrared range of the spectrum at a wavelength of 1064 nm. Therefore, in this work, we used a laser [6, 7] radiating in a shorter wavelength region of the spectrum, which contributed to an increase in the absorption capacity. A specially designed laser consisted of a synthetic alexandrite crystal (100 mm long, 6.3 mm in diameter). The ends of the crystal have an antireflection coating in the range from 700 to 820 nm. The crystal was pumped by a flash lamp with an energy from 14 to 289 J. The pump pulse repetition frequency was varied from 1 to 30 Hz. The laser operated in the free generation mode at a wavelength of 750 nm. The dependence of the output power of laser radiation on the frequency of the energy pumping pulses is shown in Figure 1.

Journal of Physics: Conference Series **2316** (2022) 012004 doi:10.1088/1742-6596/2316/1/012004

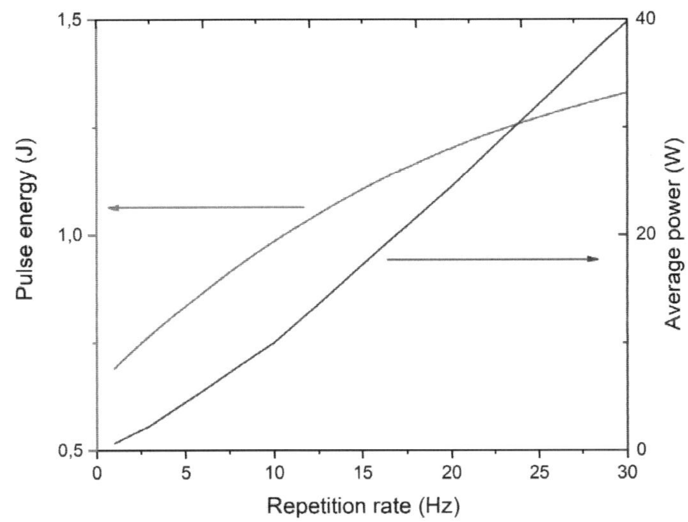

Figure 1. Dependence of the average radiation power on the repetition rate of the pump current pulses.

Figure 1 shows that by varying the parameters of the pulsed current pumping, the pulsed energy of laser radiation was changed. The formation of a laser beam on the surface of a copper target was carried out using an aspherical focusing lens. During the experiment, the energy and time parameters of laser radiation were measured. The laser pulse energy was measured using a PE-25 pyroelectric converter and Juno interface manufactured by Ophir Optronics Solutions Ltd. The wavelength and duration of the laser pulse were measured from a diffusion reflecting screen using an HR4000 High Resolution Spectrometer (Figure 2a) by Ocean Optics, Inc. and photodetector OD-08A (Figure 2b) by Avesta Project Ltd. The energy density distribution and cross-sectional diameter of the laser beam (Figure 2c) were measured using a Pyrocam IIIHR Beam Profiling Camera manufactured by Ophir Optronics Solutions Ltd.

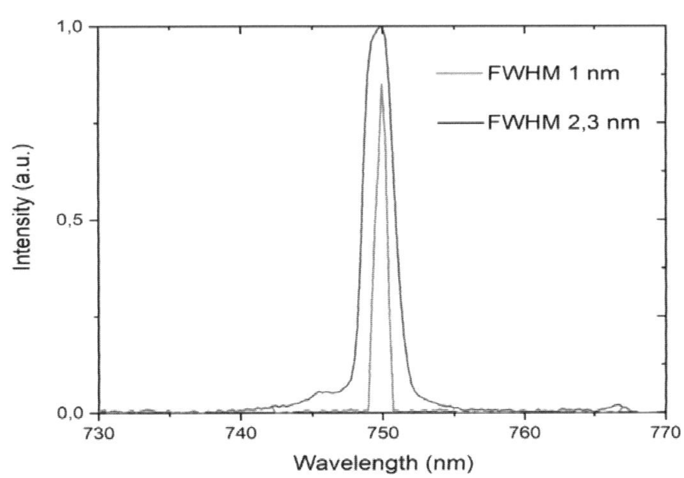

a)

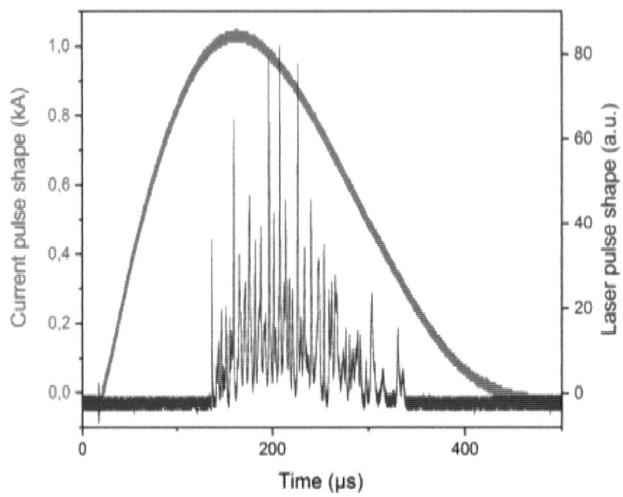

b)

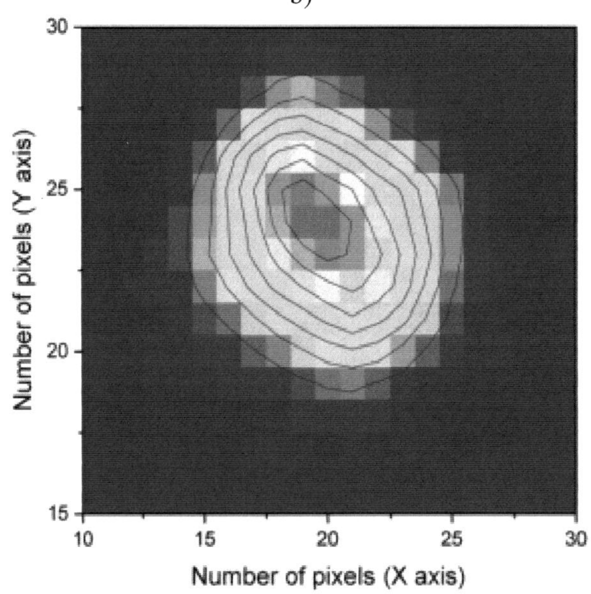

c)

Figure 2. The spectrum of laser radiation – a, the temporal shape of the laser pulse – b, the distribution of the energy density over the cross section of the laser beam – c.

3. Results and discussion

Colloidal copper solutions were formed by laser ablation of the material in a liquid. A copper target was placed in the liquid: distilled water and glycerin. The energy density of laser radiation on a copper target was about 52 J / cm^2. The result of laser action is that an air bubble is formed on the surface of the copper target with the formation of a plasma torch. This is due to local point laser heating, the beam diameter is on the order of 50-70 µm, and, as a consequence, an increase in the target temperature, which exceeds the boiling point of the liquid. Point laser action leads to surface melting of the target and rapid cooling. The formation of a surface plasma torch of various intensities occurred during different laser action times from 10 to 20 seconds. This is due to the repetition rate of the laser pulses. After the formation of the surface plasma torch, the process of laser scanning of the target began. The scanning speed of the target surface did not exceed 2 mm/s. The result of this effect is the synthesis of a one-component colloidal solution shown in Figure 3a.

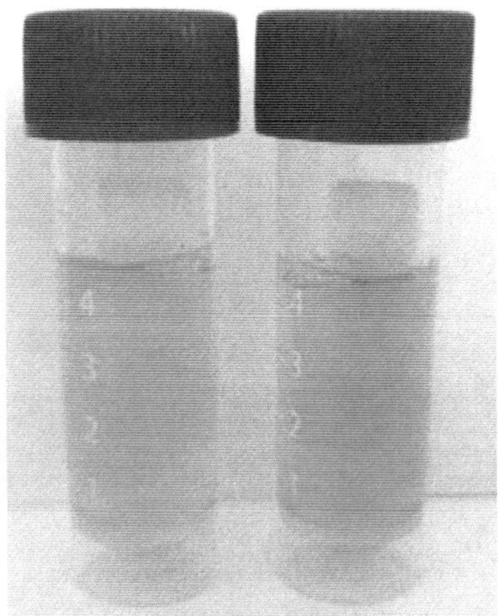

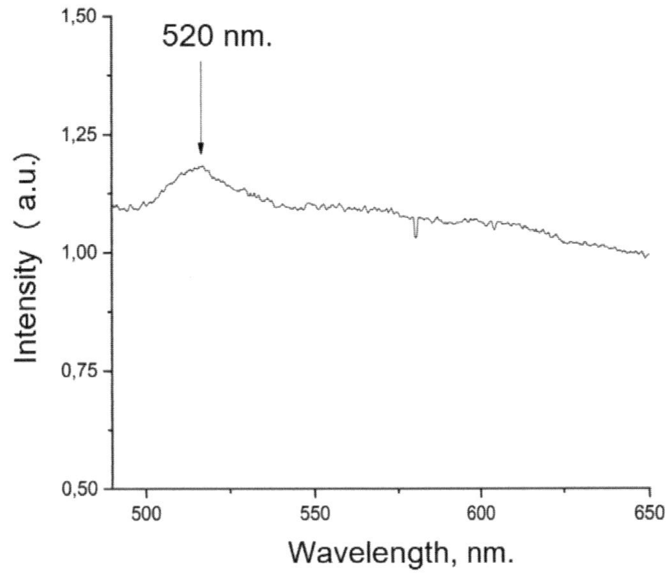

a) b)

Figure 3. One-component colloidal solutions of copper – a and absorption spectrum of the solution – b.

Absorption spectrophotometry was used to determine the absorption spectrum of colloidal solutions. For colloidal solutions with synthesized copper nanoparticles, the absorption peak was in the region of ~ 520 nm (Figure 3b). Similar absorption spectra are presented in [2, 8] in the study of colloidal solutions with copper nanoparticles obtained by laser ablation.

A Horiba LB-550 laser analyzer was used to measure the size of the synthesized particles in colloidal solution. Research on dynamic light scattering showed that nanoparticles of various sizes are formed in the process of laser ablation, depending on the experimental parameters of laser radiation. For example, when the target is laser irradiated with a frequency of 5 Hz, the average particle size is about 20 nm (Figure 4a), an increase in the laser radiation frequency to 15 Hz leads to an increase in the average particle size to 150 nm (Figure 4b). This increase is associated with a change in the profile of the laser beam, since an increase in frequency leads to an increase in the thermal lens.

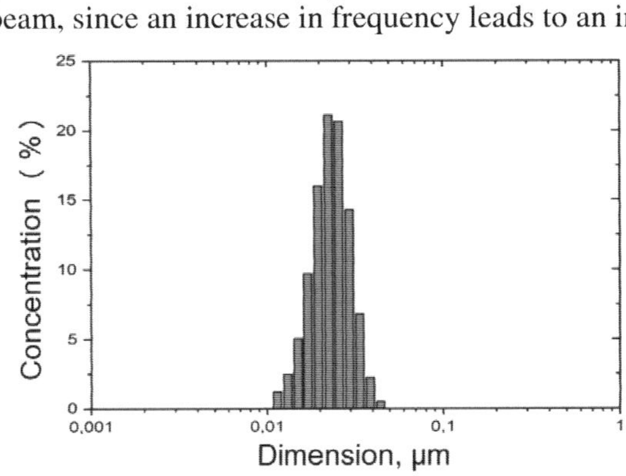

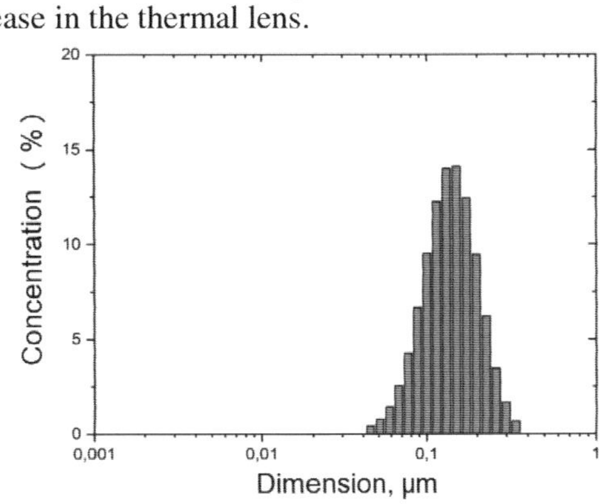

a) b)

Figure 4. Histograms of particle size distribution at a laser irradiation frequency of 5Hz-a and 15Hz – b.

A decrease in the scanning speed of laser ablation to 1 mm/s leads to an increase in the average size of nanoparticles: at a frequency of 5 Hz, the size is 150 nm and 450 nm at a frequency of 15 Hz. The change in size is associated with repeated irradiation, which leads to adhesion of particles and an oxide film forms on the surface of the particles.

With an increase in the scanning speed, while maintaining the previous laser parameters, the surface plasma torch is not formed and, as a result, the colloidal solution is not formed.

Further research is aimed at the formation of thin cluster films. Cluster films were formed on a solid dielectric substrate by drop deposition. In the first series, experiments were carried out in natural conditions at a temperature of 25 °C. The formation of a thin film and evaporation of the liquid phase was monitored using an optical microscope. In the process of drying, a cluster thin film is formed with a well-defined boundary and uniformly deposited particles in the center, Figure 5.

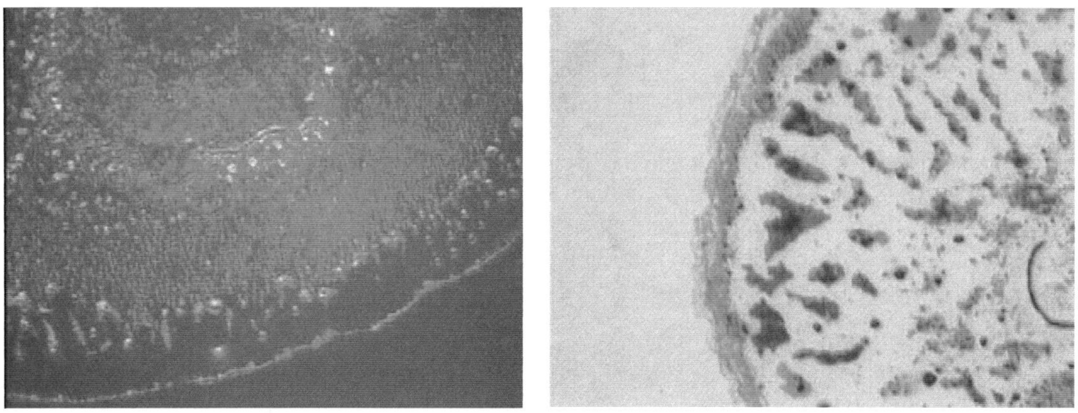

a) b)

Figure 5. Optical image of a cluster film of copper with 20X: a) natural drying b) forced drying.

The second series of experiments is devoted to stimulated evaporation of a liquid medium using laser radiation. A drop of colloidal solution was deposited on a dielectric substrate and irradiated with defocused laser radiation. The formation of a cluster film under this effect occurs in two stages. At the first stage, laser heating of the colloidal solution occurs, as a result of which its thermodynamic parameters change. This leads to a competition between two processes: first, diffusion of copper nanoparticles from the heating region to the outside, and second, hydrodynamic flow into the heating region. At the second stage, the action of surface tension forces is overcome, – the particles are deposited on the surface of the substrate with the formation of a cluster film (Figure 5b).

4. Conclusion

In the course of the work, colloidal solutions were synthesized by laser ablation. Laser ablation was carried out using the developed alexandrite laser. The investigations of the spectral and energy characteristics of laser radiation in the process of laser ablation have been carried out. It is shown that the smallest copper nanoparticles are synthesized at an energy of 52 J/cm^2, a repetition rate of 5 Hz, and a scanning speed of 2 mm/s. An increase in the laser radiation frequency leads not only to a change in the beam profile, but also to an increase in the average particle size. The formation of cluster films in the process of natural and forced evaporation of the liquid phase is demonstrated.

Acknowledgments

The reported study was funded by RFBR, project number 19-29-10022, partly project 20-32-90052, project 20-02-00515 and supported by the Ministry of Science and Higher Education within the State assignment FSRC «Crystallography and Photonics» RAS.

The study was carried out using the equipment of the interregional multispecialty and interdisciplinary center for the collective usage of promising and competitive technologies in the areas of development and application in industry/mechanical engineering of domestic achievements in the field of nanotechnology (Agreement No. 075-15-2021-692 of August 5, 2021).

References

[1] Zhu H, Zhang C, and Yin Y 2005 *Nanotechnology* **16** 3079–83
[2] Mohammed M T, Albosalih A, Saleh S M and Salih B 2019 *Kufa Journal of Engineering* **10** 1–11
[3] Fernández Arias M, Boutinguiza M, Del Val J, Covarrubias C, Bastías F, Gómez L, Maureira M, Arias González F, Riveiro A and Pou J 2020 *Appl. Surf. Sci.* **507** 145032
[4] Nag A, Frias Batista L M, Tibbetts K 2021 *Nanomater.* **11** 814
[5] Kucherik A, Ryabchikov Y, Kutrovskaya S, Al-Kattan A, Arakelyan S, Itina T and Kabashin A V 2017 *ChemPhysChem* **18** 1185–91
[6] Antipov A A, Putilov A G, Osipov A V and Shepelev A E 2020 *Bull. Russ. Acad. Sci.: Phys.* **84** 1359–61
[7] Antipov A and Putilov A 2018 *Mat. of the 7th Int. Conf. Modern nanotechnologies and nanophotonics for science and production* (08-12.11.2018 Suzdal RU) 64-65
[8] Nag A, Frias Batista L M and Tibbetts K M 2021 *Nanomater.* **11** 814

Journal of Physics: Conference Series **2316** (2022) 012005 doi:10.1088/1742-6596/2316/1/012005

Investigation of material heating under the action of ultrashort laser pulses

A V Kharkova, K S Khorkov, A V Egorova

Vladimir State University, 87 Gorky Street, Vladimir 600000, Russia

E-mail: alenaenergie@gmail.com

Abstract. The aim of this work was to study the heating of an AISI 304 plate in the ablative mode of femtosecond laser pulses. The variable parameter was the change in the speed of movement of the laser beam over the treated surface. The paper presents the results of measuring the heating temperature of the sample surface exposed to laser radiation without scanning laser radiation with different pulse energies. The results of the dependence of the heating of the material surface on the speed of movement of the laser radiation with a constant number of acting pulses are presenting.

1. Introduction

Currently, the process of laser ablation of materials when exposed to ultrashort pulses is of great scientific and practical interest [1-3]. In the process of laser ablation processing, including in liquid media, it is possible not only to process the surface of materials, but also to synthesize nanoscale materials with predictable dispersion [4, 5]. The efficiency of ablation of materials is influencing by the mode of transfer and relaxation of the received energy of laser radiation by the material, which is determining by the complex cumulative interaction of competing processes. In [6-10], a set of processes occurring during laser ablation is indicating. As the main processing parameters, it is worth highlighting the energy in the laser beam pulse, the pulse repetition rate, and the speed of the beam movement over the material surface. These parameters determine the efficiency of ablation as a whole; it is necessary to isolate the accumulation of heat by the processed material and the development of a laser erosion torch.

Ablative laser synthesis of nanoparticles makes it possible to obtain spherical granules with a narrow granulometric composition [1, 3, 11]. However, this method also has a number of disadvantages, namely, the synthesized particles can have traces of fusion, together with the main material at the outlet, traces of "splashes" can be observing, and there can be pronouncing traces of the presence of a melt pool [12-15]. This fact leads to significant contamination of the ablation products, which makes it difficult to further using this powder material. In addition to contamination, due to the melting process, an increase in the particle size distribution of the powder can also occur, which is also an undesirable effect. In order to avoid the listed negative effects, it is necessary to establish such modes in which the thermal effects will not have such a significant effect.

In [4, 16], the authors investigate the processes during femtosecond laser ablation, where it is noted that at the beginning the process of absorption of laser radiation energy occurs, which leads to the realization of the stage of melting of the surface layer of the material, and only after that the stage of hydrodynamic expansion begins. At the last stage, due to energy loss, the residual heat spreads deep into the material, which leads to its solidification (Figure 1).

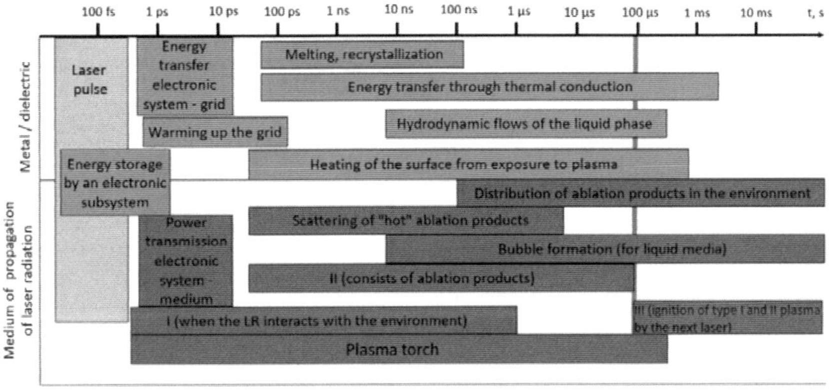

Figure 1. Diagram of the propagation of processes in time when exposed to laser radiation.

In Figure 1, the red band indicates the beginning of the next impulse. An important parameter is also the absorption capacity of the target material, as well as its thermal conductivity [16]. In the event that the material has a high absorption coefficient, then the melting stage can be more pronouncing, due to the accumulation of a large amount of energy inside the material. The same applies to thermal conductivity. Thermal processes with a high coefficient of thermal conductivity will be more pronouncing. As noted in [17], an important value is also the value of the ablation threshold, as soon as the laser radiation density in the groove is below the ablation threshold, all the absorbed energy will be converting into thermal energy. Thus, an important role in determining the processes under the action of ultrashort pulses is playing by the heat-affected zones formed during processing, which largely depend on the intensity of the action of laser radiation and laser-induced plasma [18, 19].

In this work, the problem was posing to determine the degree of thermal contribution from the regime of laser action. The change in the mode will be carrying out due to the different speed of scanning the beam over the surface with a constant number of laser pulses. The scanning speed determines the number of pulses of the laser beam exposure to the received area of the material surface. This approach makes it possible to make a predicted thermal contribution to the ablation process, since when scanning at a low speed, significant heating of the affected area is possible both as a result of heat accumulation and the thermal effect from the laser erosion torch.

2. Description and discussion of the experiment results

The experiment was carrying out in atmospheric air. The laser-induced plasma channel was not limited in space and could freely leave the affected area. The thermal contribution was measuring both in the point mode and in scanning with a laser beam. The TETA-10 femtosecond laser system with the following parameters was using as a radiation source: wavelength - 1030 nm, pulse duration - 280 fs, frequency - 10 kHz. The radiation power was varying by varying the pump current parameter and a polarization attenuator. AISI 304 stainless steel was using as a sample. The scheme of the experiment is showing in Figure 2.

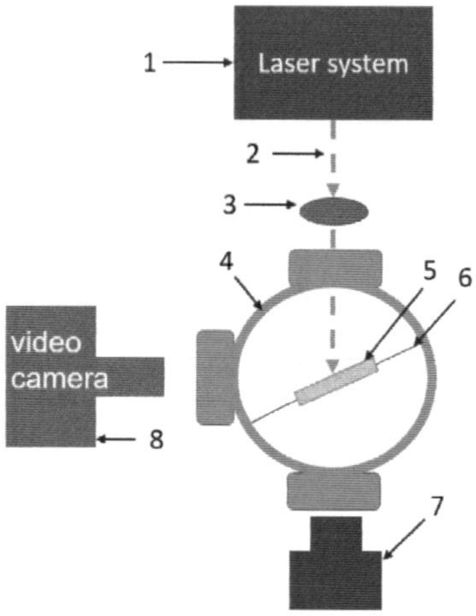

Figure 2. Scheme of the experiment.

In Figure 2 shows: 1 – laser system, 2 - laser radiation, 3 - focusing lens, 4 - camera with observation windows, 5 - sample, 6 - glass fiber sample holders, 7 - thermal imager, 8 - video camera. The laser beam had a diameter of 80 μm. The AISI 304 stainless steel sample was 10 by 15 mm in size and 500 μm thick.

The laser beam was focusing onto the sample surface using a focusing lens with a focal length of 200 mm. The video camera was focusing on the area of influence, where the presence or absence of the process of material ablation under the influence of radiation was observing. The sample holder was made of fiberglass. This holder material has been selecting to limit further heat propagation from the sample. In addition, a thermal imager was included in the scheme, with the help of which the temperature on the surface of the sample was recording. To obtain correct thermographic information, the surface of the samples facing the thermal imager was covering with a light-absorbing coating, which excluded the reflection of heat rays from the surrounding elements.

To determine the ranges of the energy density of laser radiation, a number of works were carrying out aimed at identifying the threshold of laser ablation, the energy of laser radiation leading to through penetration of the sample used. The fact of ablation was establishing based on data from the camera, by analyzing the state of the surface. The beam diameter on the substrate surface was about 40 μm. When exposed to laser radiation with an average energy density of 0,127 J/cm^2, no changes were detecting during prolonged exposure. At an energy density of 0,167 J/cm^2, pronounced changes in the surface are observing because of single pulses, we consider this energy density to be the minimum providing an ablation effect. In the interval of the indicated values, a certain darkening of the area of influence is observing during prolonged exposure. Further, the maximum energy density is establishing at which the formation of a through hole in the sample used is not observing, since when the through whole is forming, part of the radiation will pass without interacting with the substrate material. Empirically, the value of the peak energy density is 1,273 J/cm^2, which corresponds to the maximum energy value that does not lead to the formation of a through hole. The initial value of the sample temperature was 23, 5 C°. The diagram shown in Figure 3 shows the results of measurements of the heating of the plate when exposed to laser radiation with the corresponding values of energy.

Using the data obtained from a thermal imaging camera, graphs of the dependence of the heating temperature on the exposure time at different radiation power were plotting. The dependencies are showing in Figure 3.

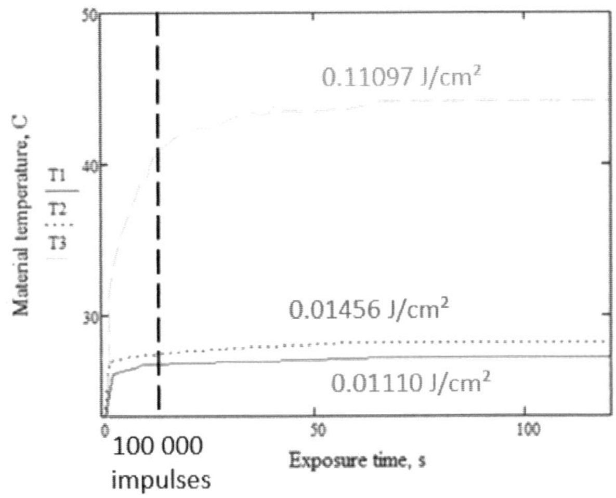

Figure 3. Dependence of heating temperature on exposure time at different radiation power.

In the graph shown in Figure 3, the red line corresponded to the radiation power of the peak energy density of 0,0111 J/cm^2. At this energy, the ablation process was not observing. After increasing the radiation power to a peak energy density of 0,01456 J/cm^2, the onset of the ablation process was recording on the camera. At a peak energy density of 0,11097 J/cm^2, effective heating of the material is observing during the ablation process. A further increase in the radiation power leads to the formation of a hole in the sample. In this regard, it is not advisable to further measure the temperature, since the radiation no longer has a significant effect on the sample. During the registration of the temperature during spot processing, the achievement of the peak heating temperature was recording, after reaching which the temperature did not increase anymore and remained within the peak values. This fact can be explaining by the fact that when processing to a point, craters were forming over time. The deeper the craters formed, the more energy had to be expending to increase the temperature of the sample, since there is an increase in the treatment area. Thus, with the effective process of ablation of the surface layer of the material, the surface is melting, in the formed cavity due to the defocusing effect, most of the laser radiation is lost, and the energy propagates over a large area, while the bulk of the energy goes into heating the material.

In the work, an experiment was also carrying out in which the measurement of the heating temperature of the material was recording when scanning with a laser beam at different speeds. The diameter of the scanning beam was 30 μm, the energy of the laser radiation was 28 J/cm^2. Scanning was carrying out using a galvanoscanator at a speed of 0,2-1000 mm/s. The total number of laser pulses directed at the sample surface was 100 000.

Figure 4 shows the dependence of the heating temperature of the material on the speed of the laser beam along the sample surface.

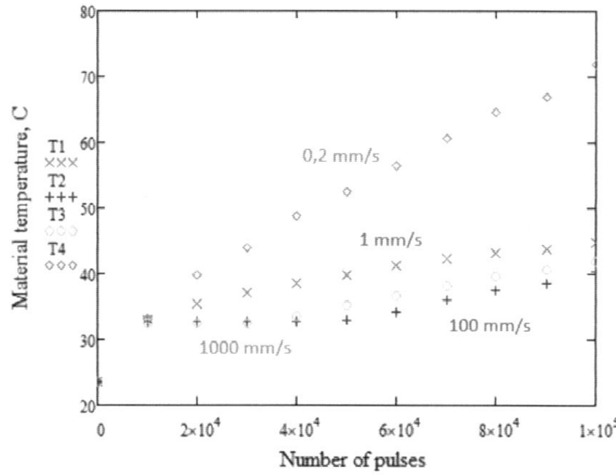

Figure 4. Heating temperature versus exposure time when scanning with a laser beam, peak fluence: 1,38713 J/cm².

Changing the speed of movement of the beam allows you to determine the degree of thermal contribution to the process of material processing. Figure 4 shows that with a decrease in the scanning speed, the heating temperature of the material increases. This can be explaining by the following: a decrease in the speed of movement of the beam increases the degree of overlap of the zones of exposure to laser radiation. It can also be noting that when scanning with a laser beam at speeds of 1-1000 mm/s, the sample is heating less intensively at a much higher laser radiation power than when processing to a point according to the data of the previous experiment with an average peak energy density of 0,0111 J/cm^2. Since the temperature reaches close values (about 40 degrees) within 10 seconds after the start of treatment (which corresponds to 100 000 pulses) This fact is due to the fact that during the scanning mode the ablation process takes place, the incoming energy is spent on removing the surface layer of the material. Thus, the treatment zone is heated and this heated material is removing from this zone. The removed material does not participate in the heat transfer processes occurring in the processed sample. When scanning at a speed of 0.2 mm / s, not only a significant heating of the sample is observing, but also a significant decrease in the ablative effect of the laser beam. Around the area of influence there is a pronounced zone of temperature influence, the resulting groove has edges that are much larger than the size of the beam, on the edges there are traces of emissions of the liquid phase of the material. In Figure 5, you can observe photographs of the material processing zone under point impact, as well as when scanning with a laser beam.

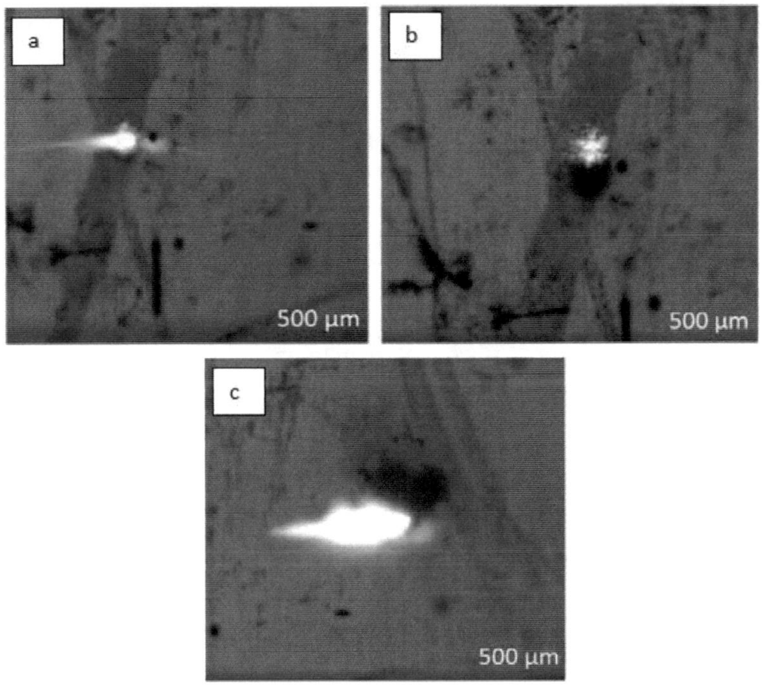

Figure 5. Images of the material processing area: a, b - with a point effect (peak fluence: 1,38713 J/cm²), under the letter c - when scanning with a laser beam (peak fluence: 1,38713 J/cm²).

In Figure 5, the processing was carrying out at peak fluence: 1,38713 J/cm², under the letters a and c one can observe the formation of a laser erosion plume, which is characterized by a similar bright glow above the surface.

When processing with radiation, the incoming energy is no longer enough to maintain the ablation process, since a crater is forming. As can be seen from [17, 20], the depth of the craters formed, as noted above, also affects the heat distribution and the nature of ablation. Laser radiation, in this case, propagates over a larger area. The radiation is defocused, which significantly reduces the effectiveness of the impact. To resume the ablation process, it will be necessary to spend more energy. Thus, almost all of the incoming energy is converting into heat (Fig. 6).

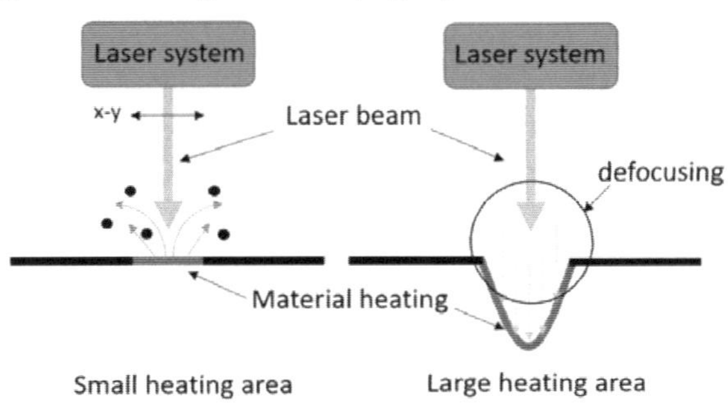

Figure 6. Exposure to laser radiation when scanning by a beam and when exposed to a point.

Figure 6 schematically shows the process of exposure to laser radiation in various modes. As can be seen from the figure, when scanning with a laser beam, the material is heating over an area approximately corresponding to the beam diameter. The escaped heated particles also freely leave the treatment area, which excludes their re-irradiation. The impact of radiation at a point leads to the formation of cavities, with an increase in their depth, the area of propagation of radiation energy also increases. The expended energy is not enough for the development of the material ablation process;

part of the radiation is lost when the laser radiation is defocused. The rest of the energy is spending on thermal processes (melting, evaporation, etc.). Thus, as noted above, all energy is converting into heat.

3. Conclusion

In this work, an experiment was carrying out to determine the heat of heating of a material during exposure to ultrashort laser radiation. It was found that a certain peak heating temperature is achieved with a point effect on the sample, above which the sample is not heating, a cavity is formed, which also does not change with further exposure. The re-investment of thermal energy due to the expansion of the ablation products that interact with the walls of the groove becomes more intense with an increase in its depth, since the area of the treated zone increases. In addition, due to the geometry of the groove, laser radiation does not affect the surface of the material at a right angle; multiple re-reflection of the laser beam with an energy not exceeding the ablation threshold leads to heating of the surface. The efficiency of the ablation process decreases with an increase in the contribution of kinetic energy to the processing area due to the contribution of energy from the laser erosion plume, which leads to the formation of a melt zone.

Thus, in order to increase the efficiency of the material ablation process, while maintaining the shape of the granules, their high smoothness, as well as a narrow grain-size distribution, scanning with a laser beam is the most effective by significantly reducing the effect of thermal processes during material processing. In the works carried out, it was establishing that at a slow scanning speed, a significant heating of the processed plate is observing. A decrease in thermal processes makes it possible to exclude the ingress of drops and splashing into the resulting material. During machining operations, reducing the thermal input allows you to increase the contrast of the edges of the resulting parts.

Acknowledgments

The study was carried out using the equipment of the interregional multispecialty and interdisciplinary center for the collective usage of promising and competitive technologies in the areas of development and application in industry/mechanical engineering of domestic achievements in the field of nanotechnology (Agreement No. 075-15-2021-692 of August 5, 2021). This work was also partially supported by the framework of the state task of VlSU № 0635-2020-0013.

References

[1] Voznesenskaya A et al 2020 *Solid State Phenomena* **299** 205–209
[2] Ionina A et al 2013 *Journal of Experimental and Theoretical Physics* **3** 403–422
[3] Kochuev DA et al 2018 *J. Phys.: Conf. Ser* **951(1)** 012015
[4] Kochuev DA et al 2018 *International Conference Laser Optics (ICLO)* 2018 335-335
[5] Chernikov AS et al 2021 *J. Phys.: Conf. Ser* **1** 012002
[6] Ionin AA et al 2017 *UFN* **2** 159–172
[7] Bulgakova NM et al 2008 *High-Power Laser Ablation VII* **7005** 70050C
[8] Kononenko TV et al 2015 *Journal of Applied Physics* **10** 103105
[9] Bulgakova N et al 2014 *Micromachines* **4** 1344-1372
[10] König J et al 2005 *Opt. Express* **26** 10597-10607
[11] Kochuev DA et al 2020 *IOP Conf. S.: Mater. Sci. and Eng.* **1** 012034
[12] Kazakevich VC et al 2018 *Brief comm. on physics FIAN* **3** 15-20
[13] Leshchik SD et al 2014 *Proc. of the Internat. Sci. and Technic. Conf* **2** 84-87
[14] Goncharova D A et al 2013 *High technol. in mod. sci. and technol* 2 136-139
[15] Noor Zirwatul Ahlam Naharuddin et al 2020 *Optical Materials Express* **2** 323-331
[16] Tsarkova OG et al 2011 *Proc.of the Institute of General Physics. A.M. Prokhorov* **67** 53-63
[17] Freitag Christian et al 2018 *Advanced Optical Technologies* **3** 157-163
[18] Chkalov R V et al 2021 *Trans Tech Public. Ltd* **316** 164-169
[19] Kononenko T V et al 2018 *Optics and Lasers in Engineering* **103** 65-70

[20] Struleva E B et al 2019 *Thermal physics of high temperatures* **4** 529-533

Laser-induced breakdown spectroscopy monitoring of aluminum alloys laser cleaning process

A A Voznesenskaya, D S Gusev, D A Kochuev, A A Polyakova and K A Frolov

Vladimir State University, Gorky Str. 87, 600000 Vladimir, Russia

E-mail: 2obk@bk.ru

Abstract. The paper considers an approach to monitoring the condition of the treated surface in real time based on the laser-induced breakdown spectroscopy technique. As a result of the study, the aluminum alloy was cleaned from the plimer coating, spectral analysis was carried out, the relationship between the processing modes and the thickness of the coating being removed was experimentally established, which allows removing the paint coating in one pass without damaging the underlying layers. The optimal spectral range for the control operation, the contrasting spectral lines of the elements that make up the coating, the optimal parameters of laser exposure are determined based on the monitoring data of the spectral signal recorded during laser processing

1. Introduction

Laser radiation is a universal technological tool that allows precise, predictable processing, including of thin-film coatings, without damaging the substrate surface [1, 2]. One of the methods of precision laser processing is laser cleaning. This method is considered a highly efficient and environmentally friendly surface treatment technology [1, 3], which does not require the use of cleaning solutions containing acids, alkalis, organic solvents. The problems of industrial use of cleaning solutions are associated with the complexity of their use, storage and subsequent disposal, belong to the II class of hazard (highly hazardous). During laser cleaning of the surface, the removed material is retained by air membrane filters, depending on the surface to be cleaned, the used filters belong to the III or IV class of hazard (moderately low hazardous). Laser cleaning of the surface makes it possible to effectively eliminate contamination of the following types: oil contamination [4–7], polymer paint coatings [8, 9] and oxide layers [10–12]. One of the main and especially urgent areas of laser cleaning of industrial facilities is cleaning the surfaces of aircraft parts. The main mechanism of paint removal was investigated in [9].

The problem of laser cleaning of the surface of aluminum alloys used as external elements of the skin of aircraft is the need to remove the paint and varnish coating without damaging the oxidized layer. Due to the large cleaning area and complex geometry, it is very difficult to ensure that the focusing system is always in the focal plane where predictable material removal occurs. Deviation from the focal plane leads to a decrease in processing efficiency. It is necessary to use fast methods of monitoring the surface condition directly during processing. The construction of such a control system is possible using the laser-induced breakdown spectroscopy (LIBS) method. When the laser radiation interacts with the surface to be treated, the emission spectra of the elements that fall into the area of exposure to the laser beam are recorded. The surface to be treated is a multilayer structure consisting of: base material (aluminum alloy), an oxidized layer - prevents material erosion as a result of the action of the atmosphere, a polymer layer - is a paint and varnish coating, renewed every 2-4 years. To ensure high accuracy when processing a surface with laser radiation, it is necessary to carry it out in several passes, making minimal material removal. Carrying out manual control after each pass is a costly and often ineffective control method, since the polymer coating is covered with a layer that has some fluctuations in thickness (from 20 to 10 micrometers). Efficient productive processing is possible by building a feedback system. In the course of processing, the emission spectra of the laser erosion plume are recorded. When the emission spectra of oxygen and aluminum are detected, the processing of the region stops; the detection of these spectra indicates the destruction of the surface of the oxidized layer. In [3], the LIBS technique was used as an element for monitoring the surface condition in real time for laser cleaning of the TiAlN coating on a WC cutting tool. The operation was monitored throughout the cleaning operation. The presence of emission spectra characteristic of titanium was a factor in the reproduction of the cycle of laser surface treatment. The transition to the next section was carried out in the absence of recorded emission spectra of Ti.

2. Experimental part

Laser cleaning of aluminum alloy B95 with AK-1206 paint-and-lacquer coating was carried out using a fiber YAG: Nd-laser, wavelength λ = 1065 nm; average power 20 W; pulse duration 100 ns, spot diameter in the working plane 35 µm, flux density 5.6 × 103 J / cm^2. Processing was carried out using a galvanometric scanner with a processing area of 100x100 mm, focusing was carried out by F-theta with a working distance of 200 mm. To record the spectral signal, an ASP-150 was used, with an adjustable slit, a grating of 1400 lines / mm; the spectral signal was fed through an optical head with an optical fiber.

Under the action of pulsed laser radiation on the treated surface in the surface layer, the laser radiation was absorbed by the coating material, the processes of evaporation, excitation, and ionization of the substance were observed. During processing, significant changes in the brightness of the glow of the laser erosion plume were visually recorded depending on the processed layer of the substrate.

The methods of subsequent control of the effectiveness of the laser cleaning operation are visual inspection and measurement of the electrical conductivity of the treated surface. By visual inspection, it is possible to reveal the presence - absence of paintwork residues. Evaluation of the electrical conductivity of the coating makes it possible to establish the integrity of the oxidized layer of the surface of the aluminum alloy. This operation was carried out using a digital multimeter CEM DT-9908. Due to the fact that the paint and varnish coating and the anodized layer do not have electrical conductivity, the absence of electrical conductivity on the treated surface was determined using a multimeter, which indicated the integrity of the oxide layer.

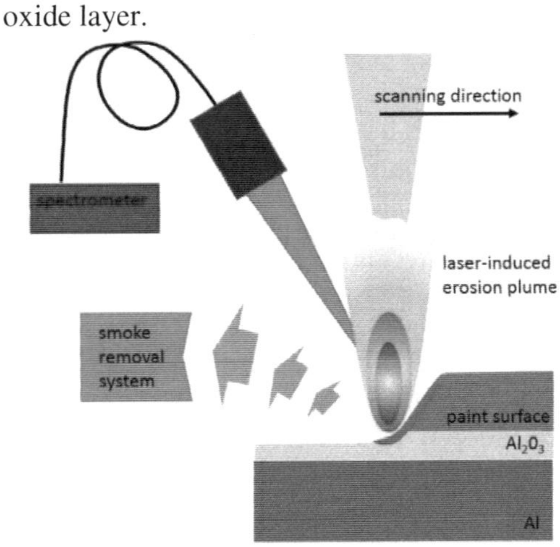

Figure 1. Diagram of the process of removing paint and varnish coating and recording a spectral signal.

To avoid the destruction of the aluminum oxide layer on the surface of the workpiece in real time, a spectral data recording system was used, which made it possible to register the spectral lines of atoms of chemical elements that make up the paint and varnish coatings, the anodized layer and the aluminum alloy. Registration of a spectral signal from a vapor-gas cloud is effective in laser micromachining in modes that exclude significant thermal effects. Exposure to laser radiation with a duration of the order of hundreds of nanoseconds makes it possible to efficiently excite atomic vibrations and obtain a contrasting line spectrum, excluding significant line broadening. The resulting sparks and flashes of materials outside the erosion torch, when ablated material or smoke enters the propagating laser beam, interfere with the recorded spectrum of Planck radiation during laser processing. To obtain spectral data from the required area of the laser erosion plume, an optical head is used, which is a 2-lens optical system with an installed aperture diaphragm and an SMA905 fiber connector. The optical head was switched to the spectrometer with a 10-meter fiber with a light-guiding core section of 400 μm.

3. Results and discussion

Since the product made of anodized aluminum alloy has a three-layer structure, the primary task was to determine the optimal range of the recorded spectral signal. To ensure the minimum hardware complexity of the detection unit, to simplify the analysis of the spectral signal, it is necessary to have spectral regions that characterize the maximum differences between the spectra of the ablated layers of materials (Figure 2):

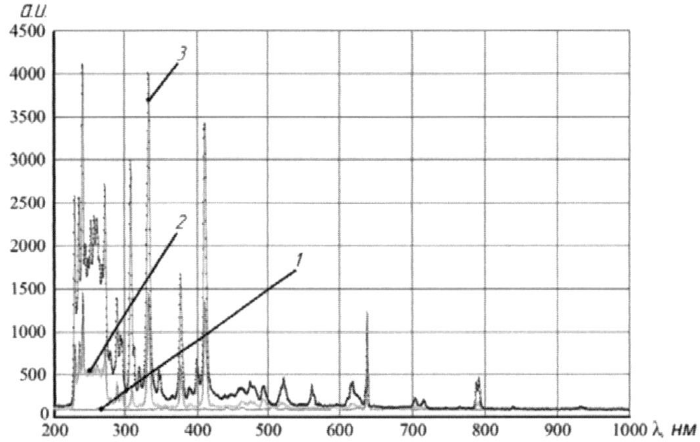

Figure 2. Comparison of spectral lines obtained with: (1) - partial removal of the paintwork, (2) - partial removal of the anodized layer, (3) - complete removal of the anodized layer.

The most suitable range was chosen in the 260-430 nm range in view of the obvious differences between the diagrams shown in Figure 2. The interpretation of the spectral lines in the indicated range when removing the paint and varnish coating is shown in Figure 3. The large width of the recorded spectral lines is caused by the maximum opening of the spectrometer entrance slit, for obtaining maximum response and minimum exposure of the matrix. This step was taken to increase the performance of the control system.

To carry out a qualitative elemental analysis, it is necessary to identify the atomic and ionic lines in the spectrum. For this, the electronic databases of spectral data NIST [13], Robert Kurucz [14], less often - Peter van Hoof [15], or tables of spectral lines [16] are used. The number of known atomic spectral lines in the listed bases reaches 140,000, 450,000 and 920,000, respectively. For complex multi-element samples (high-alloy steels, soils, ores) with a high content of transition elements (iron, chromium, titanium, etc.), the identification procedure is often difficult and takes a lot of time without using any auxiliary programs, because emission lines of various elements can overlap, superimpose, and their intensity depends on the conditions of excitation and the characteristics of the plasma. Independent, correct identification of spectral lines is a difficult and time-consuming task. The analysis of the obtained spectral information was carried out using the NIST resource in the LIBS database section [13]. In the user interface, the elements were indicated according to the percentage and content of the applied layers in the substrate material. The calculation of emission spectra in this tool is carried out using the Saha equation, which is valid for the state of thermodynamic equilibrium of the studied plasma (Saha-LTE). Evaluation of the plasma temperature and the concentration of electrons obtained as a result of laser action on the treated surface was carried out by analyzing the relative intensity of spectral lines similar to the method presented in [17-19]. The recorded spectral signals correspond to the excited state of the substance (ion I), according to the data [20]. With nanosecond ablation processing in this energy range, the recorded spectral response corresponds to a low-temperature plasma with an electron temperature of about 0.4 eV and an electron concentration of about 10^{13} - 10^{14} cm^{-3}. Deviations in the concentration of electrons, depending on the recorded signal, are explained by the different concentration and efficiency of absorption of laser radiation by the materials that make up the ablated surface. The ablative effect of the laser beam is sufficient for the formation of a laser-induced plasma above the surface of the processed material of sufficient density for a correct assessment of the spectral signal.

Figure 3 shows the emission spectra of oxygen and aluminum in an excited state to further determine the presence or absence of an anodized layer.

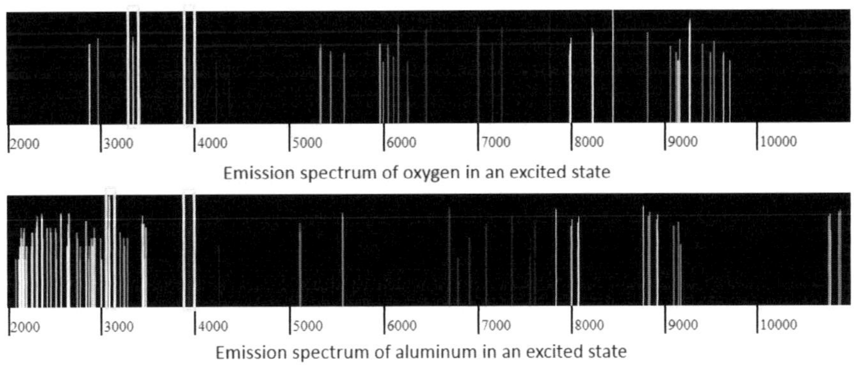

Figure 3. Emission spectra of oxygen and aluminum in an excited state [20].

The main chemical elements of the paintwork have been successfully recognized, namely Se, Fe, Na, Cr, nitroalkanes -NO_2. Unspecified intensity peaks in the range of 380-390 nm are probably spectral lines of functional groups (amino and nitro groups) containing nitrogen and oxygen, while intensities in the range of 315-320 nm and 420-430 nm are artifacts that could be on the surface of the paintwork during normal operation of the outer hull of the aircraft.

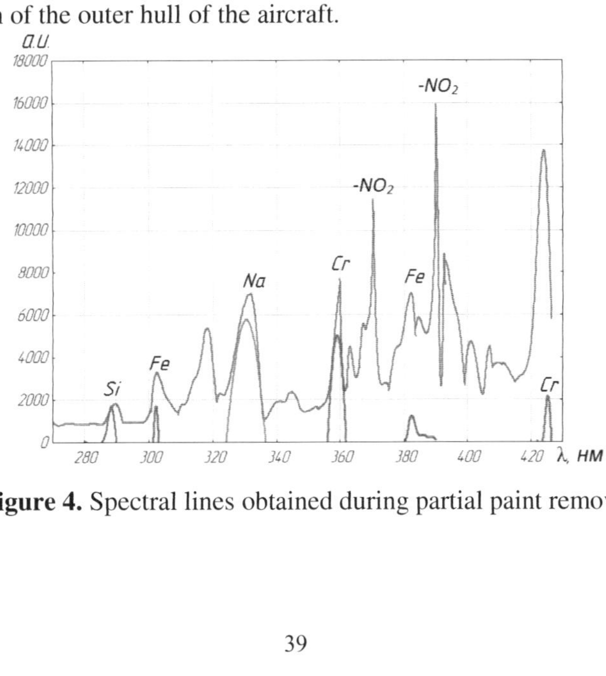

Figure 4. Spectral lines obtained during partial paint removal.

Figure 4 shows the spectral lines obtained with partial removal of the anodized layer. Only two chemical elements, Al and Cr, stand out as bright spectral lines. Other chemical elements that make up the B-95 aluminum alloy with a percentage of more than 1%, such as Mg and Cu, also have an insignificant intensity in the spectrogram.

The registration area of the spectral signal is chosen in such a way as to exclude the superposition of the spectral lines of aluminum and chromium on the spectral lines of the paint-and-lacquer coating. A group of Al and Cr spectral lines can be used as a contrast indicator of the completion of complete removal of the paint and varnish coating from the treated area.

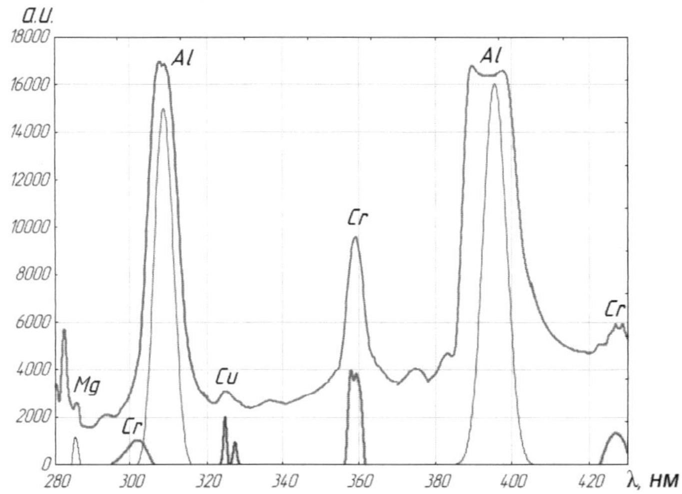

Figure 5. Spectral lines obtained during partial removal of the anodized layer.

The data obtained make it possible to formulate the basic rules for controlling the process of laser cleaning of paint and varnish coatings while preserving the anodized layer. Registration of oxygen spectra means destruction of the oxidized layer, which means that it is necessary to lower the power of laser radiation. When recording the spectra of aluminum, ablation of the base material occurs, which is unacceptable and the process must be stopped.

Figure 6 shows a cross-section of the sample after the operation of laser removal of the paint layer with process control based on the analysis of the spectra of the erosion flame obtained using optical microscopy.

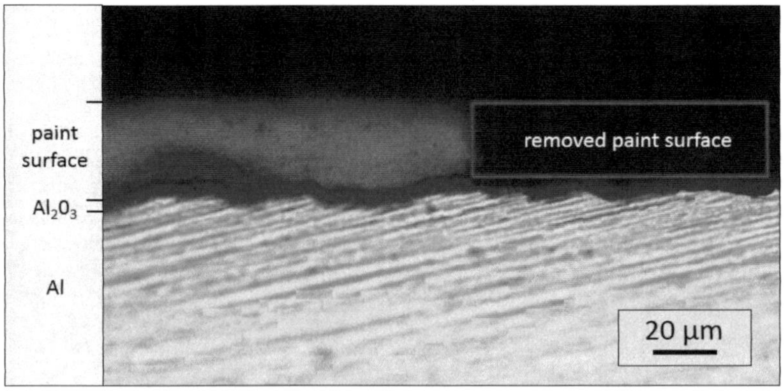

Figure 6. End section of the boundary region of the processed sample of the aircraft body skin.

On the end section of a boundary processed sample of the outer skin of an aircraft wing, containing a removable paint-and-lacquer coating based on epoxy resin, an aluminum alloy base is clearly visible;

thin black layer of oxidized coating - Al_2O_3; on the left side of Figure 6 - a layer of paint-and-varnish coating, not subject to laser treatment, on the right - the area after the laser cleaning operation.

4. Conclusion

As a result of the work carried out, the study of the spectra of the erosion torch was carried out, the surface of the aluminum alloy was cleaned, the relationship between the treatment modes and the thickness of the coating to be removed was experimentally established, allowing the paint coating to be removed in one pass to the required thickness. Based on the data of previous and current studies, the values of the electron temperature of the laser erosion torch and the concentration of free electrons were determined based on the method of estimating the relative intensity of spectral lines. The analysis of the received spectral signals was carried out using the open library NIST LIBS. An algorithm has been developed for controlling the process of laser cleaning of paint coatings by registering spectral lines Al in the wavelength range of 305-315 nm and 385-405 nm and Cr in the range of 355-365 nm when the anodized coating is reached.

Acknowledgments

The study was carried out using the equipment of the interregional multispecialty and interdisciplinary center for the collective usage of promising and competitive technologies in the areas of development and application in industry/mechanical engineering of domestic achievements in the field of nanotechnology (Agreement No. 075-15-2021-692 of August 5, 2021). This work was also partially supported by the framework of the state task of VlSU№ 0635-2020-0013.

References

[1] Chkalov R, Khorkov K and Kochuev D 2019 *Mat. Today: Proceed.* **11** 441–45
[2] Ivashchenko A V, Voznesenskaya A A , Davydov N N , Khorkov K S and Kochuev D A 2018 *J. of Phys.: Conf. Ser.* **1109(1)** 012035
[3] Veiko V, Samohvalov A and Ageev E 2013 *Opt. Laser Technol.* **54** 170–75
[4] Marimuthu S, Sezer H K and Kamara A 2019 Develop. in Surf. Contamin.and Clean.: Appl.of Clean.Tech.**11** 251–88
[5] Mateo M P, Nicolas G, Piñon V, Ramil A, Yañez A 2005 *Appl. Surf. Sci.* **247** 333–39
[6] Marimuthu S, Kamara A M, Whitehead D, Mativenga P, Li L, Yang S and Cooke K 2008 *J. Eng. Manuf.* **225** 1851–62
[7] Hu G Q, Song Y and Guan Y C 2019 *Nanotech. Precis. Eng.* **2** 29–34
[8] Zhang Z Y, Zhang, J Y, Wang Y B, Zhao S S, Lin X C and Li X Y 2018 *Optics* **174** 46–55
[9] Han J H, Cui X, Wang S, Feng G, Deng G and Hu R 2017 *J. Mod. Optic.* **64** 1947–59
[10] Zhou C, Zhou C, Li H G, Chen G Y, Wang G and Shan Z Z 2020 *Opt. Laser Technol* **127** 1061197
[11] Zhang G X, Hua X M, Huang Y, Zhang Y L, Li F and Shen C 2020 *Appl. Surf. Sci.* **506** 144666
[12] Li H, Costil S, Liao H-L Coddet C, Barnier V and Oltra R 2008 *J. Laser Appl.* **20** 12–21
[13] Kramida A, Ralchenko Yu, Reader J and NIST ASD Team 2015 *NIST atomic spectra database (ver. 5.3)* http://physics.nist.gov/asd
[14] Smith P L, Heise C, Esmond J R and Kurucz R L 2011 Atomic spectral line database, built from atomic data files from RL Kurucz
[15] Van Hoof P A M 2018 *Galaxies* **6(2)** 63
[16] Seidel A N, Prokofiev V K, Raisky S M, Glorious V A and Schrader E Ya 1969 *Spectral line tables* (Moscow: Nauka*)* p 782
[17] Tarasova M A, Kochuev D A, Galkin A F, Vasilchenkova D G, Chernikov A S, Khorkov K S and Prokoshev V G 2019 *J. of Phys.: Conf. Ser.* **1331(1)** 012018
[18] Kochuev D A, Galkin A F, Voznesenskaya A A, Khorkov K S and Chkalov R V 2020 *Bull. of the Lebedev Phys.Inst.* **47(2)** 37–42
[19] Kochuev D, Chkalov R and Chernikov A 2019 *Mat. Today: Proc.* **19** 1928–31
[20] Kazakov V V, Kazakov V G, Kovalev V S, Meshkov O I, and Yatsenko A S 2017 *Phys. Scripta* **92(10)** 105002.

Research of the physical and mechanical properties of AISI 321 steel manufactured by selective laser melting

A V Zhdanov, A S Raznoschikov, A A Voznesenskaya, K S Poltorakova and I V Belyaev

Vladimir State University, Gorky Str. 87, 600000 Vladimir, Russia

E-mail: 2obk@bk.ru

Abstract. In this paper, the problems of creating parts by selective laser melting from AISI 321 powder material are considered. Samples were made from powder material with a granulometric composition of 10-63 microns at a selective laser melting plant. Microslips were made to study the physical and mechanical properties. The structure of the samples and their porosity were evaluated using optical microscopy. The change of physical and mechanical properties depending on the change in the scanning speed is investigated. The connection between the modes of selective laser melting and the physical and mechanical properties of the resulting products has been established.

1. Introduction

At this time additive technologies are actively developing all over the world. Selective laser melting of metal powder compositions is one of the popular directions. This can be associated with the high flexibility of production (from the moment of creating a 3D model to the moment of its production, it takes a little time), the possibility of modulating the properties of the obtained one was done by changing the concentrations of various powders used. Low production waste and minimal subsequent machining are also attractive aspects of selective laser melting technology. This technology makes it possible to manufacture parts that are maximally optimized in terms of weight and strength, with complex geometry, which is not achievable with classical processing methods. One of the topical areas of application of selective laser melting is the production of unique medical products [1, 2]. In the medical industry, there is a need for structurally complex parts that work in extremely critical places and areas [3].

Despite the popularity of selective laser melting, there are a number of problems due to significant temperature gradients and suboptimal technology regimes, and the formation of defects is possible. The main defects in selective laser melting are roughness [4-6], pore formation, deformations and cracks [7-10]. The factors influencing the appearance of defects during selective laser melting are the geometric characteristics of the part [10], effective heat removal, the position of the part on the substrate, the thermophysical properties of the powder material [12, 13], the effect of the ambient temperature [14, 15].

2. Experimental work

The study of the particle size distribution of the AISI 321 powder was carried out using the Altami Studio software, which was developed for capturing, researching and processing images, as well as for carrying out measurements. Granulometric analysis showed that the powder particles are generally spherical (round), which are shown in Figure 1. There are also exceptions in the form of elliptical particles and branched particles. The particle size range varies from 0.89 to 54.8 microns. There were 480 measurements of particles in three photographs of the powder material, which were performed using a scanning electric microscope (SEM).

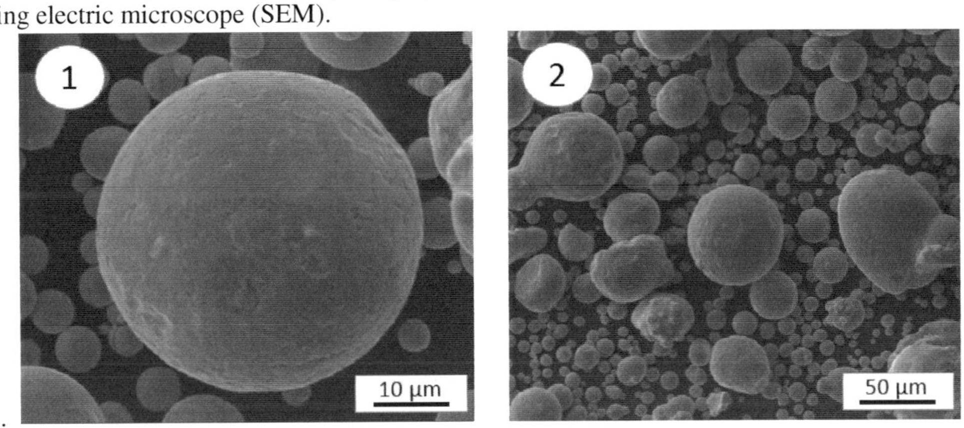

Figure 1. SEM image of powder material particles.

To study the physico-mechanical properties of the AISI 321 material obtained by the selective laser melting method, groups of primitive specimens were made on the ConceptLaser M2 Cusing industrial setup, equipped with a diode-pumped fiber laser and a beam scanning system using galvanic scanners. The maximum output power of this system is 400 W. A metal powder material 40 µm thick was deposited on the substrate surface and subjected to continuous laser action: the radiation wavelength was 1.07 µm, and the laser beam diameter on the treated surface was 120 µm. An optical system with an F-Theta lens was used for focusing. The treatment took place in a nitrogen atmosphere. The samples were made of cylindrical shape 30 mm high and 10 mm in diameter (Figure 2).

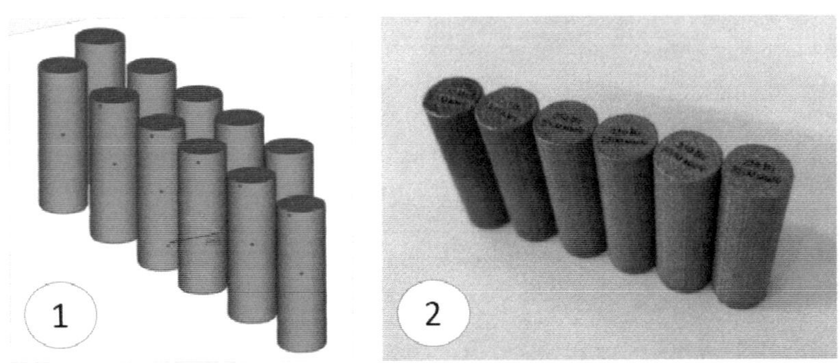

Figure 2. 1 - STL - model, 2 - samples obtained by selective melting.

Growth modes are characterized by the following parameters: laser power, scanning speed, scanning step, layer thickness. The study considered the dependence of the microstructure of the samples on the scanning speed. For the samples, the average power of laser radiation was 250 W, and the scanning speed was equal to 1000-3500 mm / s. Slice thickness 0.03mm, beam compensation 0.05mm, island dimensions X: 5mm Y: 5mm Angle 45 °. Segments of each individual layer - the so-called "islands" - are worked out sequentially. The process provides a significant reduction in stress within the component.

3. Research and discussion

Microsections of two surfaces were prepared: the surface facing the base of the construction slab (growth plane) and the end surface of the sample. The study of the porosity value from the energy input into the selective laser melting process was carried out by calculating the ratio of the total number of image pixels to pixels corresponding to the color and brightness of the pore region. The results of metallographic examination of the samples are shown in Figure 3. All images of microsections were obtained with an Altami optical microscope.

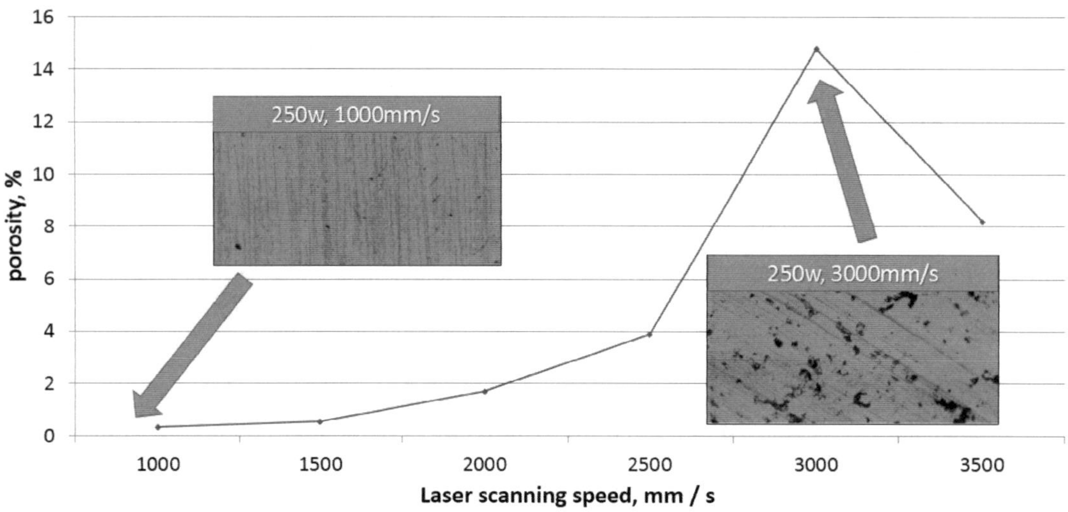

Figure 3. Dependence of porosity on laser scanning speed.

It can be seen that with an increase in the scanning speed, the porosity of the sample increases. The graph of porosity change versus scan rate is also shown in Figure 3.

Hardness was measured using a Rockwell stationary hardness tester. The results of measuring the hardness of the samples under study are shown in Table 1.

Table 1. Hardness measurement

Laser scanning speed, mm / s	Hardness, HRC					Average hardness, HRC
1000	30	34	29	35	31	31,8
1500	31,5	31	31	30,5	30	30,8
2000	31,5	28	29	28,5	28	29
2500	23	21,5	19	26	23	22,5
3000	20	17	19	18,5	19	18,7
3500	less 17					

It was found that with an increase in the scanning speed, the hardness of the samples under study decreases. The maximum value of hardness was obtained on a sample made at a scanning speed of 1000 mm / s, the minimum - on a sample made at a scanning speed of 3500 mm / s.

Compressive strength was studied using a WDW-1000 setup. The test results are presented in the form of a graph of the dependence of the load on the absolute deformation (shortening) of the sample (Figure 4).

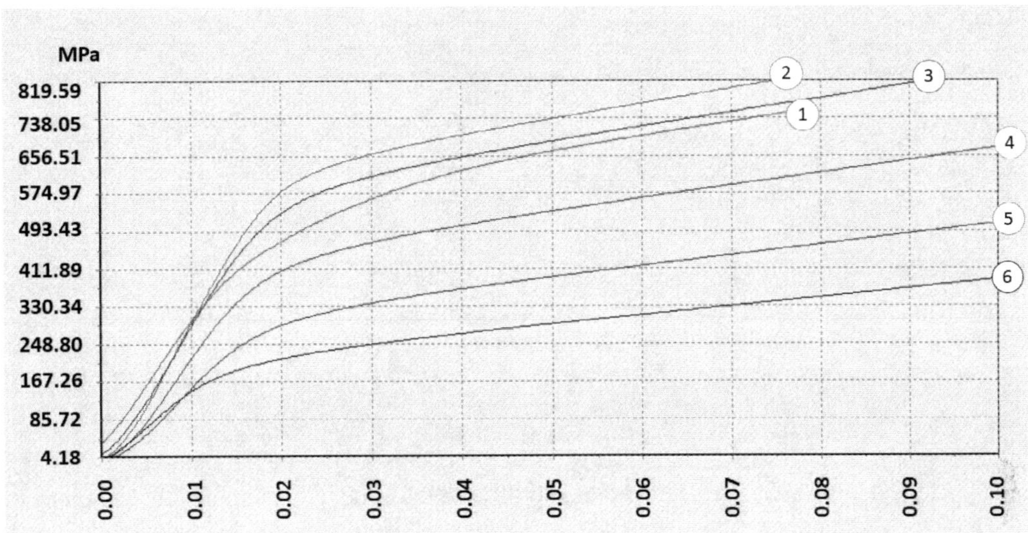

Figure 4. Graph of the dependence of the load on the absolute deformation of the samples.

In appearance, the compression graph of the samples, as for most plastic materials, approximately to the yield point coincides with the tensile diagram. The point of the diagram corresponds to the proportionality limit of the material, after which a small area is marked, where a faster growth of deformations is observed. Subsequently, the curve goes up sharply due to the increase in the cross-sectional area of the sample and the hardening of the material.

The diagram shows: line 1 - 1000 mm / s, line 2 - 1500 mm / s, line 3 - 2000 mm / s, line 4 - 2500 mm / s, line 5 - 3000 mm / s, line 6 - 3500 mm / with. The samples are sufficiently plastic, therefore, no destruction occurred, only deformation occurred. Based on the data obtained from the compression tests, it is possible to determine the maximum stress that occurs in the samples due to the applied load. For 1000 mm / s - 600 MPa; 1500 mm / s - 660 MPa; 2000 mm / s - 600 MPa; 2500 mm / s - 470 MPa; 3000 mm / s - 330 MPa; 3500 mm / s - 250 MPa.

The study of the samples for the values of friction and wear was carried out using a CSM Instruments tribometer. During the test, the samples were subjected to mechanical stress with the following parameters: normal load - 10 N, 1/2 amplitude - 6 mm, distance traveled - 200 m, maximum linear velocity 10 cm / s.

The main characteristics that were measured: penetration depth of the indenter, coefficient of friction, wear distance. To determine the average value of the coefficient of friction, it is necessary to plot the dependence of the distance on the coefficient of friction at each moment of time. The dependence graph is shown in Figure 5.

When processing the data, it is necessary to take into account two values: 1-the coefficient of friction in the running-in area, which is responsible for the initial values when the indenter is introduced into the surface of the sample; 2-coefficient of friction in the area of steady-state wear, showing normalized wear. The dependence of the wear resistance of the material on the increase in the laser scanning speed of the sample surface is shown in Figure 5.

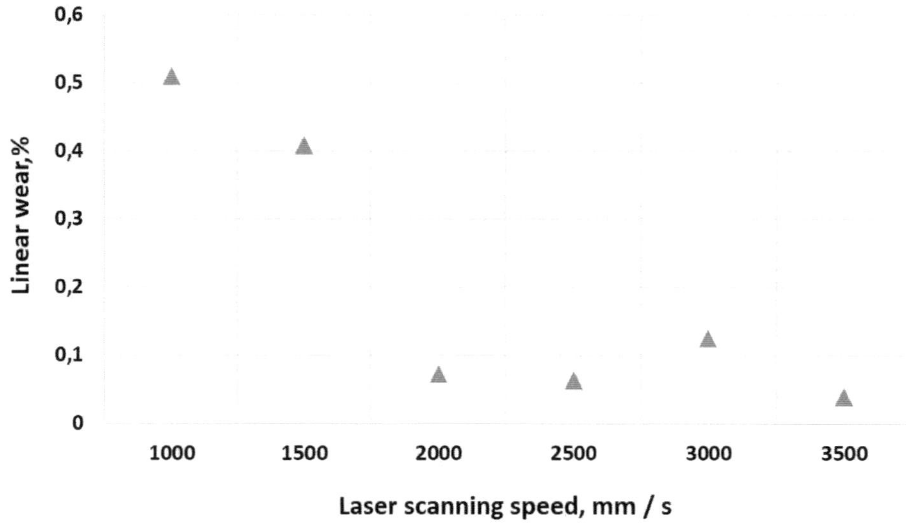

Figure 5. Graph of the dependence of the wear resistance of the material on the increase in the scanning speed of the laser on the surface of the samples.

When testing for friction and wear, it was found that the amount of linear wear decreases with an increase in the laser scanning speed of the sample surface, and the coefficient of friction increases.

4. Conclusion

In the study of the physical and mechanical characteristics, it was found that the hardness of the samples decreases with an increase in the laser scanning speed, which can be associated both with an increase in the surface porosity and with the phase composition. When testing for friction and wear, it was found that the amount of linear wear decreases with increasing laser scanning speed over the surface of the powder material, and the coefficient of friction increases. When measuring the compressive strength of the samples, the maximum values of the stress arising in the samples due to the applied load were determined. For 1000 mm / s - 600 MPa; 1500 mm / s - 660 MPa; 2000 mm / s - 600 MPa; 2500 mm / s - 470 MPa; 3000 mm / s - 330 MPa; 3500 mm / s - 250 MPa. Metallographic analysis established that the porosity of the samples increases with an increase in the laser scanning speed. Based on the study, it was concluded that the optimal modes of the laser scanning speed for a power of 250W are 1000, 1500 and 2000 mm / s.

Acknowledgments

This work was carried out as part of the RFBR project №20-08-00310 and also partially supported by the framework of the state task of VlSU № 0635-2020-0013.

The study was carried out using the equipment of the interregional multispecialty and interdisciplinary center for the collective usage of promising and competitive technologies in the areas of development and application in industry/mechanical engineering of domestic achievements in the field of nanotechnology (Agreement No. 075-15-2021-692 of August 5, 2021).

References

[1] Farber E, Nazarov D, Mitrofanov I, Sufiiarov V, Popovich A and Maximov M 2020 *Mat. Today: Proc.* **30** 746–51

[2] Kong D, Ni X, Dong C, Lei X, Zhang L, Man C, Yao J, Cheng X and X Li 2018 *Mat. & Des.* **152** 88–101

[3] Van der Stok J, Van der Jagt O P, Amin Yavari S, De Haas M F, Waarsing J H, Jahr H, Van

Lieshout E E M, Patka P, Verhaar J A, Zadpoor A A and Weinans H 2013 *J. of Orthop. Res.* **31(5)** 792-99
[4] Sachdeva A, Singh S and Sharma V S 2013 *The Int. J. of Adv. Manufact. Techn.* **64(9-12)** 1505–16
[5] Raznoschikov A S, Kochuyev D A, Voznesenskaya A A, Oparin E S and Khorkov K S 2020 *IOP Conf. Ser.: Mat. Sci. and Engin.* **896(1)** 012130
[6] Cherry J A, Davies H M, Mehmood S, Lavery N P, Brown S G R and Sienz J 2015 *The Int. J. of Adv.Manufact.Techn.* **76(5-8)** 869–79
[7] Voznesenskaya A A, Kochuev DA., Raznoschikov A S and Chkalov R V 2020 *IOP Conf. Ser.: Mat. Sci. and Engin.* **971(2)** 022004
[8] Kruth J P, Deckers J, Yasa E and Wauthle R 2012 *Proc.of the inst. of mech. engin., Part B: J. of Engin. Manuf.* **226(6)** 980–91
[9] Casavola C, Campanelli S L and Pappalettere C 2009 *The J. of Strain Analysis for Engin. Des.* **44(1)** 93-104
[10] Hagedorn-Hansen D, Bezuidenhout M, Dimitrov D and Oosthuizen T 2017 *South Afr. J. of Industr. Engin.* **28(3)** 200–12
[11] Voznesenskaya A A, Kochuyev D A, Raznoschikov A S, Zhdanov A V and Morozov V V 2020 *IOP Conf. Ser.: Mater. Sci. and Engin.* **896(1)** 012131
[12] Voznesenskaya A, Kochuev D, Zhdanov A, Morozov V and Khorkov K 2018 *J. of Phys.: Conf. Ser.* **1109(1)** 012056
[13] Melnikova M A, Kolchanov D S and Melnikova D M 2017 *Photonics* **2** 42–49
[14] Kempen K, Thijs L, Vrancken B, Buls S, Van Humbeeck J and Kruth J-P 2013 *Solid Freef. Fabric.Symp. Proc.*131–39
[15] Zhang B, Dembinski L and Coddet C 2013 *Mat. Sci. and Eng.: A.* **584** 21–31

Birefringent tuner for narrowband alexandrite laser

A G Putilov[1, 2], A A Antipov [1, 2], A E Shepelev[1], A V Osipov [1, 2] and S M Arakelian[2]

[1]ILIT RAS — Branch of FSRC "Crystallography and Photonics" RAS, 1 Svyatoozerskaya, Shatura, 140700, Moscow region, Russia
[2] Vladimir State University, Gorky Str. 87, 600000 Vladimir, Russia

E-mail: putilov.iplit@yandex.ru

Abstract. This article discusses the possibility of using Lyot interference-polarization tuners to create tunable lasers. The use of various variations and layouts of phase-shifting plates makes it possible to generate both dual-wave and single-wave radiation with the possibility of tuning. The tuning range depends both on the specific type of tuner used and on the parameters of the cavity and active medium. Dual-wave and single-wavelength generation of an alexandrite laser is demonstrated.

1. Introduction

Tunable solid-state lasers are widely used in the field of remote sensing of the atmosphere, laser spectroscopy, and hyperpolarization of nuclear spins. [1, 2]. Active media with a wide gain band, such as LISAF, Ti:Sapphire, and Alexandrite, are used to create high-power tunable lasers in the visible and near-IR ranges.

The Lyot tuner belongs to the interference-polarizing filters. Such filters are used as selectors of laser radiation and preselectors of narrow-band and single-frequency lasers. [3] When used as a preselector, another dispersion element is placed in the laser cavity. One of the simplest optical schemes of a cavity with an electro-optical filter and a Lyot preselector is shown in Fig. 1. In an electro-optical crystal, transmitted radiation is subjected to the combined effect of natural and induced birefringence. The ratio of the actions depends on the orientation of the crystal and the level of the applied field. In such cavity, the radiation is tuned by changing the strength of the field applied to the electro-optical crystal. Electro-optical filters have a high tuning rate. In terms of selectivity, electro-optical filters are significantly inferior to Lyot tuners. Construction of multi-element electro-optical filters is possible according to Wood's scheme, while the sharpness factor of such a filter will be relatively low.

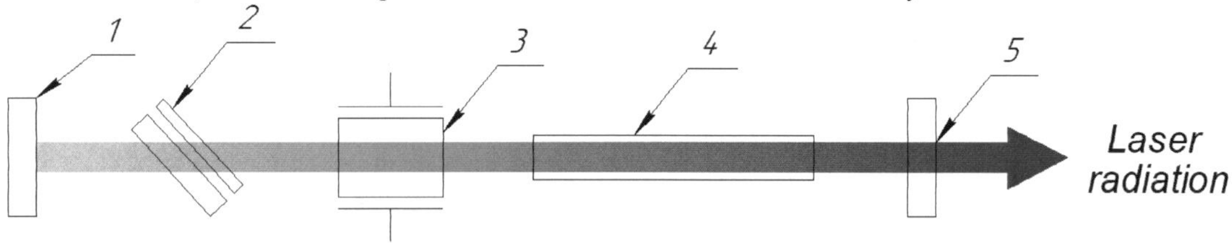

1, 5 – cavity mirrors, 2 – Lyot preselector, 3 – electro-optical filter, 4 – active media

Figure 1. Schematic of a laser with an electro-optical filter and a Lyot preselector.

A significant narrowing of the peak transmission with a simultaneous expansion of the free dispersion region is achieved in Lyot multi-element tuners. Such a tuner is a set of birefringent plates. Plates made of crystalline quartz of various thicknesses are used. The transmission of a tuner consisting of one birefringent plate is described as the difference in phase incursion between ordinary and extraordinary waves. (1)

$$T = cos^2\pi(n_e - n_o)l_c\nu \quad (1)$$

l_c – birefringent plate thickness, ν – laser emission spectrum.

Most often, in multistage Lyot tuners, phase-shifting plates of thickness are used, which form a geometric progression with an index of 2. The plates are oriented at a 45 ° angle. However, the use of plates that do not satisfy the above condition makes it possible to create dual-wavelength lasers with a terahertz frequency shift. This is achieved by installing a plate with a phase shift of π / 2 relative to other plates, the transfer function of such a plate has a minimum exactly at the wavelength at which the other filter plates show a maximum. [3] Figure 2 shows the transmission of a four-component Lyot tuner with a thickness ratio of 1: 2: 4: 8 and a total thickness of 6 mm.The resulting Lyot tuner transmission is determined as the product of the transmission of all elements (2).

$$T_r = T_1 \cdot T_2 \cdot \ldots \cdot T_n \quad (2)$$

Figure 3 shows the transmission of the same filter, but the optical axis of the thickest plate is rotated by π / 2. In this case, the resulting transmission will have two peaks. In both figures, the resulting transmission is highlighted with a black solid line. The use of such a tuner in a laser cavity will provide dual-wavelength generation with the same power at each wavelength. This becomes possible provided that the nonselective losses in the cavity for these wavelengths are equal.

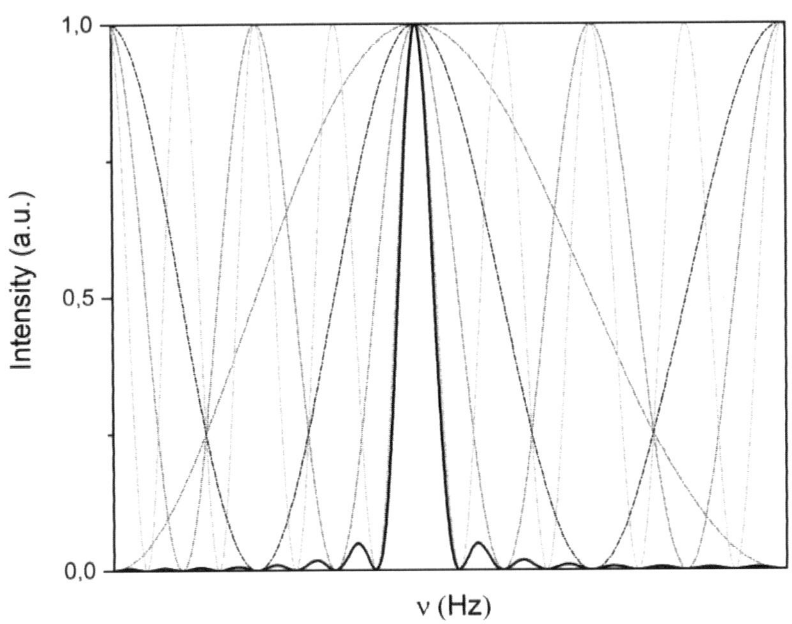

Figure 2. Transmission of a four-component Lyot tuner.

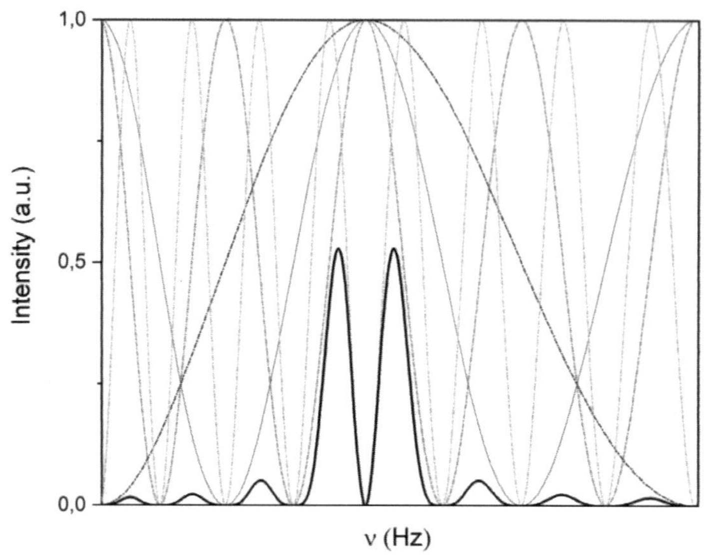

Figure 3. Transmission of a four-component Lyot tuner when the phase of the thickest plate is $\pi/2$ relative to the other plates of the tuner.

If we remove a plate with a thickness of 1.6 mm from the arrangement of this filter, then the resulting transmission of the tuner will take the form that combines both conditions for the selection of both one wavelength and dual-wave generation. (Figure 4).

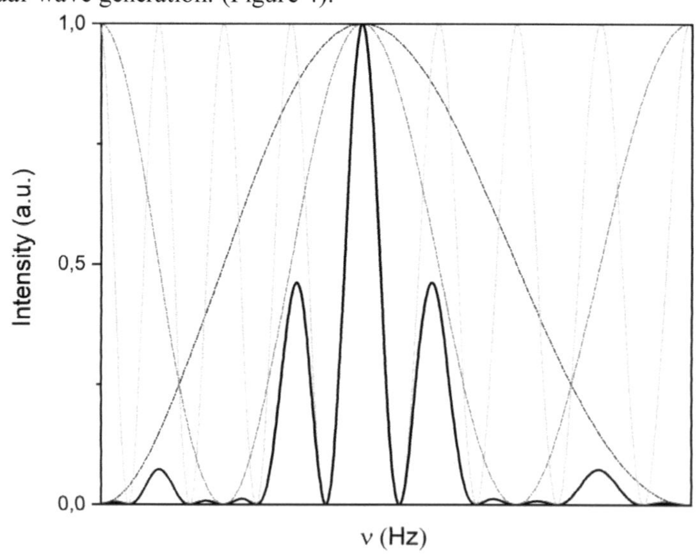

Figure 4. Three-component tuner transmission.

2. Results and discussion

To investigation the dynamics of tuning and the possibility of obtaining dual-wavelength lasing, we designed an Alexandrite laser. A Lyot tuner consisting of three phase-shifting plates was installed in the laser cavity at the Brewster angle. The center wavelength of the lasing at room temperature was 750 nm. The laser was tuned by rotating the tuner elements around the optical axis. At certain positions of the tuner, conditions for stable dual-wave lasing were created. With further rotation of the tuner, the tuning occurred abruptly, the spectrum took the form of single-wave lasing and was tuned in the range from 715 to 792 nm. Figure 4 shows single and dual-wavelength generation of an Alexandrite laser.

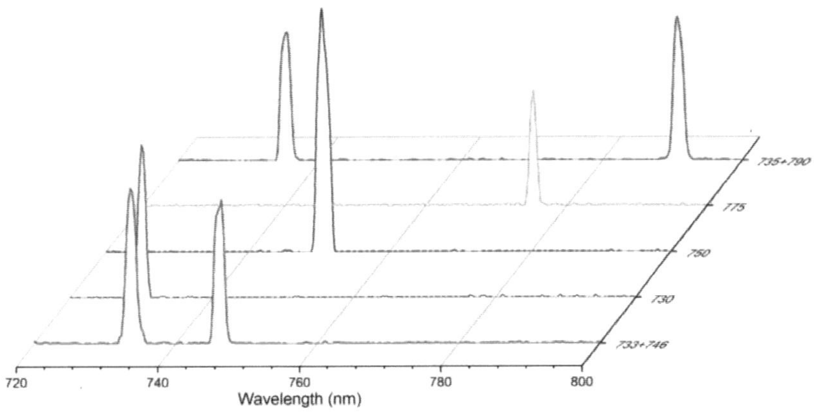

Figure 5. Tuning spectrum and dual-wavelength generation of an alexandrite laser.

Acknowledgments

The reported study was funded by RFBR, project number 19-29-10022, partly project 20-32-90052, project 20-02-00515 and supported by the Ministry of Science and Higher Education within the State assignment FSRC «Crystallography and Photonics» RAS.

The study was carried out using the equipment of the interregional multispecialty and interdisciplinary center for the collective usage of promising and competitive technologies in the areas of development and application in industry/mechanical engineering of domestic achievements in the field of nanotechnology (Agreement No. 075-15-2021-692 of August 5, 2021).

References

[1] Strotkamp M, Munk A, Jungbluth B and Hoffmann H-D 2019 *CEAS Space J* **11** 413–422
[2] Antipov A A, Putilov A G, Osipov AV and Shepelev A E 2020 *Bull. Russ. Acad. Sci. Phys.* **84** 1359–61
[3] Putilov A G, Antipov A A, Shepelev A E, Lotin A A, Arakelyan S M 2019 *J. Phys.: Conf. Ser.* **1331** 012016
[4] Treviño-Palacios C G, Zapata-Nava O J, Mejia-Uriarte E V, Qureshi N, Paz-Martinez G, Kolokolstev O 2013 *J. Europ. Opt. Soc. Rap. Public.* **8** 13021
[5] Putilov A G, Antipov A A, Shepelev A E, Arakelyan S M 2021 *J. Phys.: Conf. Ser.* **1822** 012016

Laser synthesis of plasmonic gold nanoparticles

V P Veiko, A A Morozova, D S Lutoshina, N N Shchedrina, T A Nguyen, M V Lonshakova, M M Sergeev and G V Odintsova

ITMO University, 49 Kronverksky Pr., St. Petersburg 197101, Russia

E-mail: gvodintsova@itmo.ru

Abstract. This work investigates the results of the laser irradiation synthesis of nanoparticles on the gold surface. The sizes and chemical composition of produced nanoparticles were determined, the possibility of changing the colorimetric characteristics of the gold surface was shown.

1. Введение

Since ancient times, precious metals have appealed to human attention due to their appearance and noble properties. Gold has always been distinguished among other precious metals due to its unique bright color. Because of this as well as other qualities, such as chemical and mechanical resistance, this noble metal is widely used in jewelry making.

One of the principal features of the jewelry industry is giving an aesthetic and artistic appearance to a product, which, in particular, can be achieved by changing its color.

There are numerous methods of coloring precious metals such as oxidation [1,2], galvanic method [3], applying enamel [4], etc. However, they are not universal, have restrictions on the color palette size and resolution of the applied image, and imply the necessity of consumables.

The usage of laser irradiation for coloring precious metals makes it possible to apply color images with high accuracy in a contactless way. These distinctive properties provide high-performance technology and broad prospects for implementation in production. Previously, the possibility of coloring gold by forming plasmonic nanoparticles on its surface using picosecond laser irradiation has been shown. Purple, blue and green colors, which were independent of the angle of light incidence, were created on the gold plate [5].

However, nanosecond lasers have a lower cost and higher pulse energy and offer much faster processing velocities, which makes them more preferred for industry usage, compared to picosecond lasers. In our previous works the formation of plasmon colors on the silver surface using a nanosecond laser source have been shown [6, 7]. Currently, developing technology for coloring the gold surface using nanosecond laser irradiation is of interest.

2. Material and methods

A 24K gold plate with 0.3 mm thickness was used for experiments. At the stage of preparation, a gold plate was polished with a diamond suspension (Ra = 0.769 μm). Then the samples were washed in an ultrasonic bath with deionized water.

Laser exposure was carried out in air using a system based on a ytterbium fiber pulsed laser with a wavelength λ = 1.06 μm (Laser Center Co., Ltd., Russia). The irradiation was carried out in the air in a scanning mode with overlapping pulses along one coordinate axis.

The laser generates pulses of τ = 4-200 ns duration at repetition rate in the range of f = 20–99 kHz. The beam diameter in focus is d_0 = 50 μm. Laser fluences ranged between 3.8 and 25.7 mJ/cm^2.

Visual characterisation of the obtained structures was carried out using a Zeiss Axio Imager A1M optical microscope in the reflection mode. The morphology and elemental composition of the obtained structures were detected using a Zeiss Merlin (SEM) scanning electron microscope with an Oxford Instruments INCA X-Act attachment for X-ray microanalysis. Photometric studies were carried out in reflected light using a microscope-spectrophotometer MSFU-K.

3. Results and discussion

Scanning lines were obtained on the gold plate with laser irradiation (Fig. 1). As can be seen from the figure, the sample processing area can be divided into several zones. In the first zone, direct laser action was performed with the highest processing temperature. In the second sector (about 20 μm from the center of the groove) a color change can be noticed. In the third zone (more than 40 μm from the center of the groove), where no laser processing was performed, a gradient changing color can be seen. The reflection spectra of these zones clearly show the shift of the minima of the reflection spectra to the infrared range (from zone 2 to 6), and the further shift of the minima occurs to the ultraviolet region (zone 7), approaching the reflection spectrum of unirradiated gold (Au).

Also, in the third zone, the concentration and diameter of nanoparticles decrease with distance from the laser treatment zone (Fig. 2). At the same time, in the place of the highest concentration of particles, agglomeration (aggregates) of small particles into larger particles is observed.

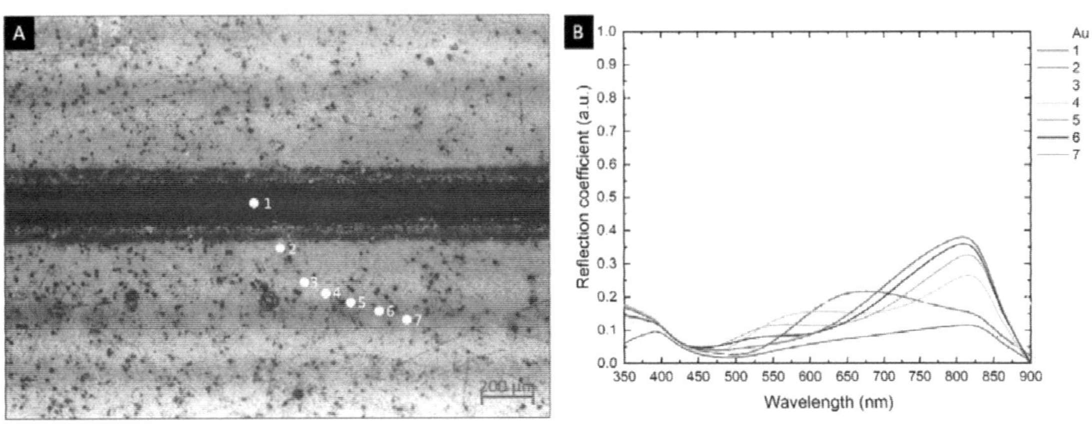

Fig.1 Microimage of laser treated area, numbers (from 1 to 7) show zones for which reflection spectra were measured (A). Color characterization of the sample: reflection spectra of the laser processed zones (from 1 to 7) on gold surface (B).

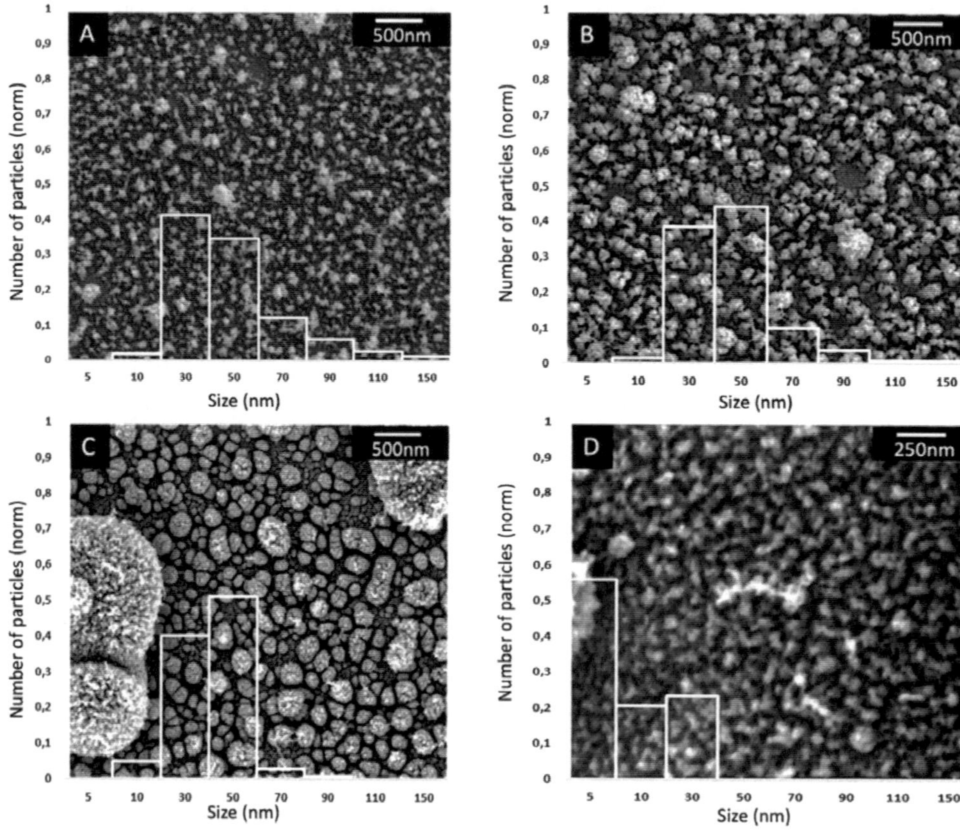

Fig.2. The morphology of obtained colors on the sample zones 6(A), 4(B), 2 (C), 1(D): SEM images and size distribution histograms. Average particle size for samples: d(6) = 38±7 nm, d(4) = 36±6 nm, d(2) = 29±9 nm, d(1) = 6±3 nm.

An X-ray dispersive analysis of the processed surface showed that the nanoparticle layer consists of gold with an insignificant presence of oxygen and carbon in the laser-irradiated region (up to 5%). This may indicate the adsorption of organic molecules from the environment by gold nanoparticles, which leads to the decay of plasmonic properties over time.

4. Conclusions

In this work, we demonstrated the coloration of a gold plate surface using a nanosecond fiber laser source for synthesizing nanoparticles with plasmonic properties on its surface.

The reflection spectra of the obtained nanoparticles were obtained, the distributions of the sizes of the nanoparticles in the colored areas were determined using SEM analysis, and the chemical composition was acquired using X-ray dispersive analysis.

Also, it was shown that the size and distribution of particles depends on the dispersal of nanoparticles from the treatment area. This will allow more precise control over the laser pulse overlapping parameters for recording multiple lines, which will make it possible to record color images on the gold surface for industrial use.

Table 1. Energy Dispersive X-Ray Analysis (EDX) of samples.

Number of zone	O (%)	C (%)	Au (%)
1	4,04	2,06	93,90
2	1,86	4,96	93,18
3	4,82	1,8	93,39
4	1,52	3,13	95,36
5	1,73	1,6	96,02
6	1,65	2,15	96,16
7	-	1,92	98,08

5.Acknowledgments

The reported study was funded by RSF, project number 19-79-10208.

References:

1. Bartlett L. 2006 An unusual phenomenon observed when anodising CP titanium to produce coloured surfaces for jewellery and other decorative uses *Optics & Laser Technology* **38**.4-6 440-444
2. D. Duprez and F. Cavani 2014 Handbook of Advanced Methods and processes in oxidation catalysis: from laboratory to industry *World Scientific*
3. H. Murray 1995 Insoluble anodes for precious metal plating *Met. Finish* 93-94
4. J. Werge-Hartley 2010 Enamelling on precious metals *Crowood Press, Limited*
5. Guay J. M. et al. 2017 Laser-induced plasmonic colours on metals *Nature communications* **8**.1 1-12
6. Andreeva Y. M. et al. 2019 Laser coloration of metals in visual art and design *Optical Materials Express* **9**.3 1310-1319
7. Odintsova G. V. et al. 2019 High-resolution large-scale plasmonic laser color printing for jewelry applications *Optics express* **27**.3 3672-3681

Er^{3+} doped sesquioxide Y_2O_3 optical ceramics: synthesis and spectral-luminescent properties

V Balashov[1], R Chkalov[2], A Fedin[2,4], M Gerke[2], K Gorbachenya[3], V Kisel'[3], D Kochuev[2], S Kozlova[1], N Kuleshov[3] and K Lopukhin[1]

[1]Kotelnikov Institute of Radioengineering and Electronics of Russian Academy of Sciences, Mokhovaya 11-7, Moscow, 125009, Russia
[2]Vladimir State University, 87 Gorky Street, Vladimir, 600000, Russia
[3]Belarusian National Technical University, 65 Nezavisimosty Ave., Minsk, 220013, Belarus
[4]Kovrov State Technological Academy, 19 Mayakovsky, Kovrov, 601910, Russia

E-mail: miron_gerke@mail.ru

Abstract. In this conribution we describe the technological process of transparent Er^{3+}:Y_2O_3 ceramic synthesis, a prospective medium for future laser applications. Hot isostatic pressing (HIP) was used to improve the optical quality of the final product. As the results of our efforts the material with an optical transmission of 0.78-0.80 was obtained, that is close to the theoretical maximum value of 0.83. The main spectral and luminescent properties of the created Er^{3+}:Y_2O_3 optical ceramics in the spectral range of 1.4-1.65 microns was determined and is reported in this work.

1. Introduction

Manufacturing and usage of optically transparent ceramics is currently one of the interesting and unresolved problems in photonics. This type of materials has great prospects for application in various fields of optics and the laser technology, since its mechanical and optical properties are superior to those of a traditional material such as glass. Dopation with ions of rare earth elements (neodymium Nd, erbium Er, ytterbium Yb and others) makes it possible to use ceramics as active laser elements.

One of the main materials for the ceramic elements production are yttrium Y_2O_3, lutetium Lu_2O_3 and scandium Sc_2O_3 sesquioxides [1]. An important advantage of laser media based on them is their 1.5-2 times higher thermooptical and thermomechanical characteristics as compared to the traditional active element based on the $Y_3Al_5O_{12}$ (YAG) single crystal [2,3].

To date, the main erbium doped materials which provide effective generation in 1.55 μm spectral range are phosphate glasses, characterized by low thermal conductivity (~ 0.85 W/m K). This fact significantly limits their use in high-power lasers. Single crystals that are widely used in laser technology, such as oxides YAG, YVO_4, fluorides LiSAF, YLF etc., although easily doped with erbium, they are effective only in a four-level scheme at a wavelength of about 3 μm (transition $^4I_{11/2} \rightarrow {}^4I_{13/2}$). In crystalline media, highly efficient generation in the region of 1.5÷1.6 microns is possible only with resonant laser pumping in the 1.5 microns spectral range [4,5].

The spectral, luminescent and generation characteristics of Nd^{3+}:Y_2O_3 ceramics have been studied before [6]. To a lesser extent, the spectroscopic and laser characteristics of this ceramic activated by Er^{3+} ions has been covered.

In this paper, the synthesis, spectral and luminescent properties of optical sesquioxide ceramics Y_2O_3 doped with Er^{3+} ions are investigated in needed detail.

2. Ceramic fabrication

The synthesis of optical ceramics was carried out in three stages:

- preparation of base powders and sintering additives;
- preparation of high-density samples from them (compacts with a density higher than 50% compared to the corresponding crystals);
- preparation of nonporous optical ceramics from the compacted base.

The method of reactive sintering was chosen as the basis technology for producing the ceramics, where the synthesis of Er^{3+}:Y_2O_3 from the initial oxides of Y_2O_3 and Er_2O_3 occurs during sintering. For the manufacture of ceramic samples Er^{3+}:Y_2O_3 (0.5 at.%, 0.25 at.% erbium respectively) commercially available yttrium and erbium oxides powders (purity worse than 99.99%) are utilized. Since a very high temperature, more than 1950 °C, is needed to obtain a transparent material (i.e., with a relative density of more than 99.95%), La_2O_3 (0.5 at%) is added to the composition to reduce the sintering temperature. However, lanthanum oxide promotes the growth of crystallite grains, so it is necessary to add components that limit this process. For these purposes, a combination of ZrO_2 (about 1.5 at%) and MgO oxides was added to the composition. MgO was introduced in the form of a $Mg(NO_3)_2$ solution in isopropyl alcohol to increase the accuracy of the suspension and a more uniform distribution of the additive. When heated, $Mg(NO_3)_2$ decomposes to MgO. The initial oxide powders and sintering additives were weighed with an accuracy of at least $2 \cdot 10^{-2}$ %.

The powders were placed in a plastic container, into which ZrO_2 balls (volume fraction 4.0 relative to the powder) and anhydrous isopropyl alcohol (volume fraction 6.0) were also loaded. The oxides of sintering additives were introduced directly into the container with the exception of zirconium oxide, which was introduced as a result of grinding ZrO_2 balls. The amount of introduced zirconium oxide was determined by the results of weighing the balls before and after milling. The milling process was carried out on a planetary mill of the Netzsch PM 400 type. Total milling time is 15 hours.

At the end of the grinding, the container with the balls was kept in a drying cabinet at a temperature of 60 °C for 30 hours. The balls were separated from the powder, and the powder was rubbed through a 240 mesh nylon sieve. The resulting powder was annealed in an alundum container at a temperature of 800 °C for 5 hours so that the powder mixture from the state with two possible structures passed into a single-phase state, since at a temperature above 750 °C the monoclinic phase turns into a cubic phase [7], this procedure was also helpful in removing any organic inclusions.

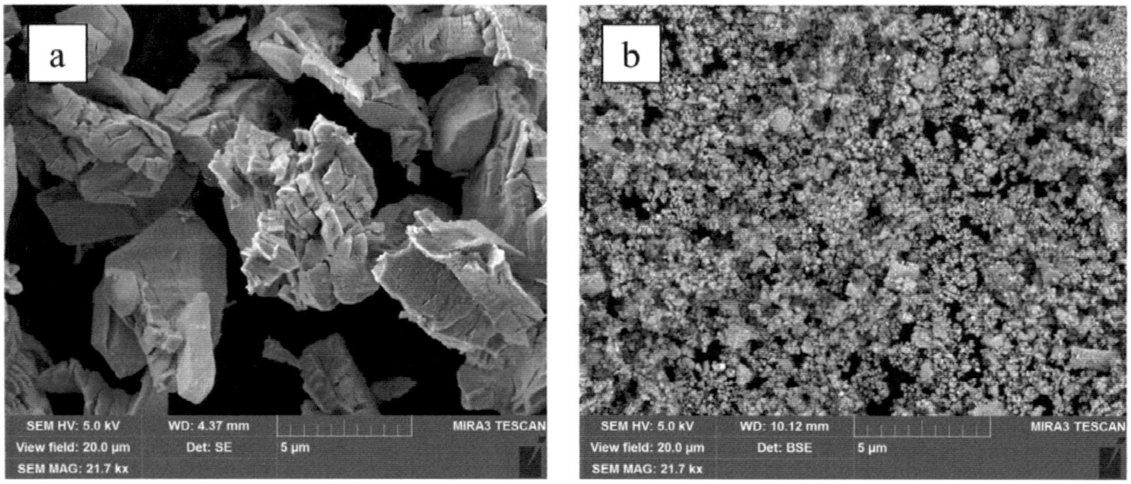

Figure 1. – SEM images of Y_2O_3 powder before (a) and after milling (b).

From Figure 1 it can be seen that proposed milling mode makes it possible to obtain a uniform fraction of Y_2O_3 powder from the initial powder with a larger grain size of tens of micrometers.

As the next step, compacts were prepared from the obtained powders. Compacts were prepared in two steps by pressing dry powder by the uniaxial compression in a metal mold. After pressing, the compacts were annealed in air at a temperature of 700 °C for 15 hours to remove volatile impurities. Then the initial samples were hermetically packed and re-pressed using a Cold Isostatic Pressing (CIP) in oil at a pressure of 250 MPa with exposure at maximum pressure for 5 minutes. After pressing, the compacts were annealed at a temperature of 1050 °C for 10 hours to finally remove all volatile impurities and eliminate residual stresses.

The resulting compacts were sintered for 20 hours in vacuum at a temperature of 1600-1700 °C until complete shrinkage with a relative density of 97-99% and then pressed in a high-pressure vessel of a hot Isostatic Pressing (HIP) press. Preliminary sintering at the temperature of conventional sintering was not carried out, since (based on our previous experience) grain growth to a size of 7-15 microns was observed. To realize the advantages of the HIP method, sintering under pressure should be subjected to non-t fully sintered ceramics with an average grain size in the range of 0.8-3.0 µm.

Hot isostatic pressing was carried out in an high purity argon medium at a temperature of 1700°C and a pressure of 2000 bar. The time cyclogram of the process was as follows:

1. Temperature and pressure (from the initial pressure of 400 bar):
   - rise from 20 to 1100 °C at a speed of 540° C/hour;
   - rise from 1100 to 1400°C at a speed of 360°C/hour;
   - rise from 1400 to 1700 °C at a speed of 240 ° C/hour.
2. Exposure: 2000 bar, 1700 °C, 4 hours.
3. Cooling at a speed of 600 °C/hour.

The heating mode with decreasing speed was applied on the basis of dilatometric measurements in order to reduce the disequilibrium of the processes occurring in ceramics and leading, among other things, to the formation of large pores and cracks.

It should be noted that deviation from the described mode of ceramics synthesis Er^{3+}:Y_2O_3 led to negative results. In the case of pre-sintering, in particular at a temperature of 1575 °C, before hot isostatic pressing due to the formation of large pores and cracks, we obtained complete destruction of samples. In the absence of annealing between pre-sintering and HIP due to the formation of large grains up to 15 µm in size samples have low density and optical transparency.

3. Ceramics characterization and optical properties measurements

XRD analysis was used to determine the phase composition of the ceramics samples. It can be seen that there are only peaks corresponding to the Y_2O_3 cubic crystal lattice. No other phases were detected, which is explained by the absence or small concentrations of sintering additives in the resulting material.

Figure 2 shows a photograph of a synthesized sample of Er^{3+}:Y_2O_3 optical ceramics after finishing polishing. The transmission of synthesized samples at a wavelength of 1550 nm was 0.8 (erbium doping 0.5 at.%) and 0.78 (erbium doping 0.25 at%), which is close to the theoretical transmission limit of 0.83 in the infrared range of the spectrum. Subsequent measurements were carried out using polished plane-parallel plates with a thickness of 2.0 mm.

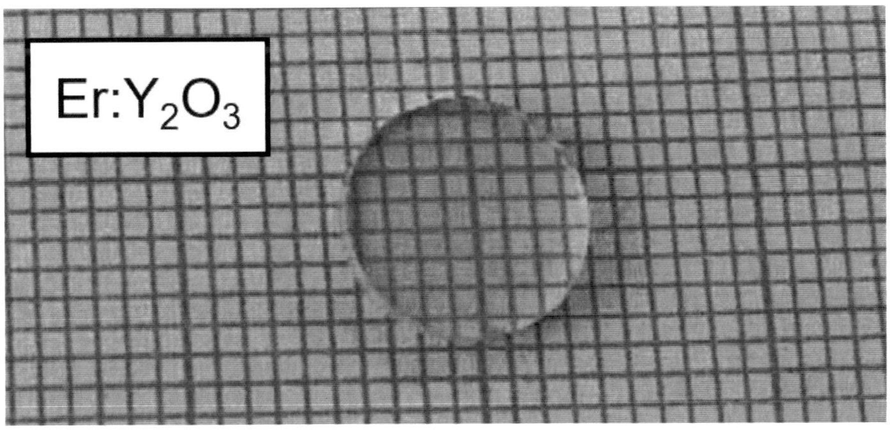

Figure 2. – Er:Y_2O_3 ceramic sample after polishing.

The average grain size was no more than 3 μm, which corresponds to the expected value for ceramics after HIP and can be reduced with further optimization of synthesis modes. Note that ceramics produced without HIP technology usually have a grain size of 5-10 μm or more [6], which reduces its microhardness and resistance to cracking.

At first the spectral transmission ∝ (λ) of Er^{3+}:Y_2O_3 ceramic was measured using Varian CARY 5000 spectrophotometer in 1400÷1650 nm range with 0.5 nm resolution. The Fresnel reflection from sample surfaces was taking into account for subsequent calculations. Then absorption cross section $\sigma_{abs}(\lambda)$ was calculated as

$$\sigma_{abs} = \frac{\alpha(\lambda)}{N}, \quad (1)$$

where λ – hereinafter light wavelength; N is Er^{3+} ions concentration, $1.34 \cdot 10^{20}$ cm^{-3} or 0.5 at.% in our case.

Figure 3 below shows the absorption cross section of the ceramic sample:

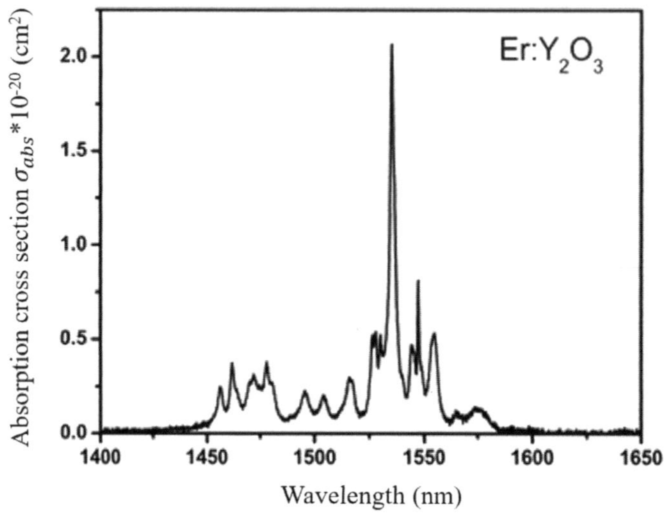

Figure 3. The absorption cross section spectra of Er^{3+}:Y_2O_3 ceramic sample.

The maximum absorption cross section was $2.0 \cdot 10^{-20}$ cm^2 at 1535 nm. The value at a wavelength of 1531 nm at which our pump source lasing and which is planned to be used in laser experiments for resonant pump of erbium ceramics is $0.42 \cdot 10^{-20}$ cm^2.

The luminescence kinetics of excited states of Er^{3+} ions was measured using the self-made setup. OPO LOTIS TII LT-2214 with tuning range 715÷2300 nm was used as a source of luminescence excitation. Luminescence was excited at 1530 nm. The luminescence decay signal was registered at a wavelength of 1580 nm with MDR-12 monochromator and InGaAs photodiode Hamamatsu Photonics G5854. The final picture showing the luminescence decay curve presented in Figure 4:

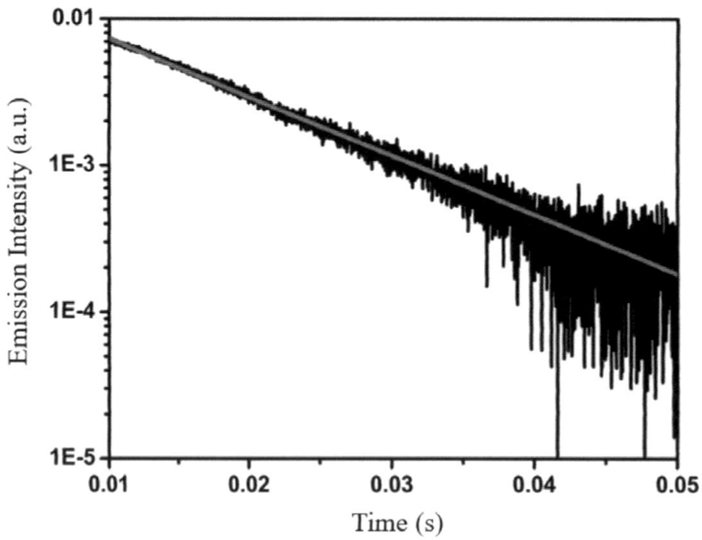

Figure 4. Time dependence of the luminescence signal.

The emission kinetics was single exponential, with the decay time measured at the level e^{-1} to be equal to 7.5±0.5 ms.

If we assume that the luminescence quantum yield at the $^4I_{13/2}\rightarrow^4I_{15/2}$ transition of Er3+ ions is close to unity, the luminescence decay time obtained as a result of measurements can be used as an estimation of the radiation lifetime τ of the $^4I_{13/2}$ energy level [5,8].

The stimulated emission cross section spectra $\sigma_{em}(\lambda)$ of Er3+ in region about 1.5 μm (transition $^4I_{13/2}\rightarrow^4I_{15/2}$) calculation was carried out according to (2) with previously obtained absorption cross sections $\sigma_{abs}(\lambda)$ and the radiative lifetime τ:

$$\sigma_{em}^{\alpha}(\lambda)=\frac{3\exp\left(-\frac{hc}{kT\lambda}\right)}{8\pi n^2\tau c\sum_{\beta}\lambda^{-4}\sigma_{abs}^{\beta}(\lambda)\exp\left(-\frac{hc}{kT\lambda}\right)d\lambda}\sigma_{abs}^{\alpha}(\lambda), \quad (2)$$

where h and k – Plank and Boltzmann constants respectively; T – medium's temperature; n – refractive index.

In the general case of an anisotropic medium such as birefringent single crustal, summation is performed over all possible polarization states. In our case, the medium is isotropic, parameters σ_{abs} and σ_{em} don't depend on the polarization state and we get a single term. The resulting curve is demonstrated below (Figure 5):

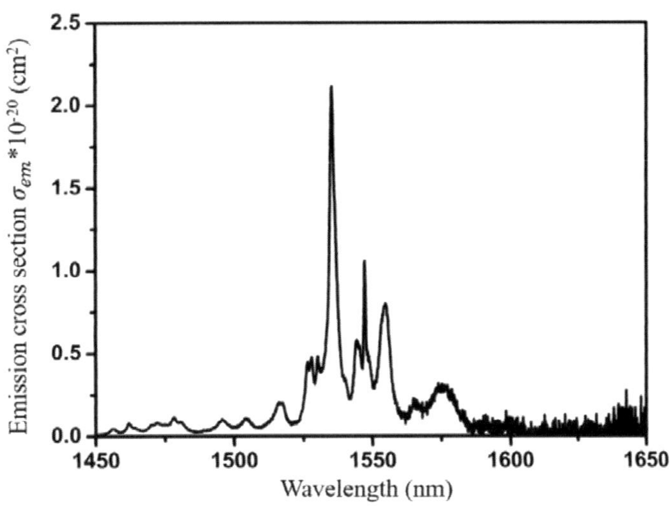

Figure 5. Emission cross section σ_{em} of Er:Y_2O_3 ceramic.

The largest maximum of σ_{em} parameter is equal to $2.1\cdot10^{-20}$ cm^2 and is observed at a wavelength 1535 nm. The cross section of stimulated emission at a wavelength of 1575 nm on which it is expected to get a laser effect in the future is $0.28\cdot10^{-20}$ cm^2.

4. Conclusion

We have described synthesis technology for Y_2O_3 sesquioxide ceramics fabricated by vacuum sintering and HIP process.

The main spectroscopic characteristics have been determined. It was found that the maximum absorption cross section is $2.0\cdot10^{-20}$ cm^2 at a 1535 nm wavelength. The cross section of stimulated emission at a wavelength of 1575 nm is $0.28\cdot10^{-20}$ cm^2. The lifetime of the excited state of the $^4I_{13/2}$ level of erbium ions was determined to be 7.5 ± 0.5 ms.

Obtained results indicate that Er^{3+} doped Y_2O_3 optical ceramics are promising laser medium. We are also in the process of developing other sesquioxides such as Lu_2O_3 and Sc_2O_3, which may provide improved thermal conductivity versus traditional YAG. Our major goal for the future work is to develop such materials that are suitable for solid-state lasing around 1.0 and 1.5 micron wavelengths.

Acknowledgments
This work supported by the RFBR project № 20-52-00015.

References

[1] Liu Z., Ikesue A., Li J. 2021 *J. Eur.Ceram. Soc.* **41** 3895
[2] Kaminskii A.A., Akchurin M.S., Gainutdinov R. et al. 2005 *Crystallogr. Rep.* **50** 869
[3] Ivanov M.G.,Kopylov Y.L., Kravchenko V.B. et al. *Inorganic materials* 2014 **50**. 1
[4] Yang X.F., Shen D. Y., Zhao T. et al. *Laser Physics* 2011 **21**. 1
[5] Gorbachenya K.N., Kurilchik S.V., Kisel V.E. et al. *Quantum Electron.* 2016 **46** 95
[6] Bagaev S.N., Osipov V.V., Ivanov M.G., *Quantum Electronics* 2008 **38** 40
[7] Groza J.R. *NanoStructured Materials* 1999. **12** 987
[8] Ter-Gabrielyan N., Fromzel V., Ryba-Romanowski W. et al. *Opt. Lett.* 2012 **37** 1151.

Deflection of particles of a laser erosion plume under the action of a high-voltage electrostatic field during ablation processing of titanium

I V Firsov, A N Zolotov, I M Efremov and A V Egorova

Vladimir State University, Gorky Str. 87, 600000 Vladimir, Russia

E-mail: firsoff2010@rambler.ru

Abstract. The ablative particles of the laser erosion torch during the processing of titanium with subpicosecond pulses were analyzed for vertical deflection under the action of an electrostatic field. The installation of a solid-state ytterbium laser with a radiation wavelength of 1030 nm was used. The photographs of the registered deviations of ablation particles in the region of propagation of the electrostatic field were obtained. The dependence of the vertical deflections of ablated particles on the laser pulse energy and the distance from the electrodes to the target is calculated.

1. Introduction

Laser ablation of materials in a medium is a complex physical process that depends on the parameters of laser radiation, the properties of the working medium, and additional sources of physical impact. An additional source of influence can be both the thermal load caused by the action of the laser-induced plasma torch [1], which is formed above the treated surface, and the energy applied to the treatment area together with the laser beam [2-5]. The combination of various methods of exposure allows achieving a synergistic effect, and, as a consequence, increasing the efficiency of the ablation operation. An effective method of combined action in the tasks of laser processing of materials is the use of an electric field [6-8]. The electric field can facilitate the removal of ablation products from the region of laser beam propagation [9-14] or for the formation of an electric arc stabilized by a laser-induced plasma channel (LIPC), which is formed when intense subpicosecond laser pulses pass through the medium in the caustic region of the laser beam [15, 16]

The use of combined exposure when performing various types of material processing, including the use of laser radiation, is a common practice [17, 18]. Usually, laser radiation acts as the main physical impact, additional impacts greatly contribute to an increase in the processing efficiency. In [19] and [4], the influence of a high-voltage pulsed electrostatic precipitator on the efficiency of deposition of nanoparticles from the gas flow of a metallurgical enterprise is considered. In [9], the efficiency of electrostatic deposition of micro- and nanoparticles in a wide range of sizes was estimated. The study of the influence of the parameters of the electrostatic field, its polarity and potential difference on the efficiency of the capture of various fractions of particles has been carried out. High efficiency of capturing micro- and nanoparticles (more than 99%) at a sufficiently high gas flow rate is noted. The use of electrostatic filters for the deposition of fines from gas streams is an effective solution.

2. Experimental part

In this work, we analyzed the deflection of titanium particles in the volume of a laser erosion plume during laser ablation and an acting electrostatic field of 15 kV/cm. The experiment was carried out in several stages, in which the parameters of laser radiation and the distance between high-voltage electrodes were changed. A femtosecond solid-state laser system TETA-10 based on an ytterbium crystal with a maximum radiation power of 10 W, a pulse duration of about 300 fs, and a wavelength of 1034 nm was used in the experiment. To register the deviations of the erosion plume, a PointGrey FL3-U3-32S2C-CS CCD camera was used, which has a maximum frame recording rate of 60 fps and a resolution of 1920 x 1080 pixels. The variable focal length micro lens used has several adjustable lenses for precise focusing. The microlens must provide flexible focusing on the target with the ability to change the angle of inclination. For effective registration of deviations, the lines of force of the external electrostatic field should be located perpendicular to the axis of propagation of the laser erosion plume. To create an electrostatic field with a strength of 15 kV/cm in the working medium of the laser, a monoblock high voltage source with voltage regulation at the output contacts is included in the circuit. The voltage source contacts were connected to fixed stainless steel electrodes. As deflecting electrodes, we used rectangular plates with a thickness of 1 mm with sharp edges and an elongated rod with a thickness of 0.8 mm, made of stainless steel, located on a movable vertical precision translator relative to the irradiated target. The type of electrodes in the form of rectangular plates has a higher probability of air breakdown to the target and subsequent closure of high-voltage contacts. The distance between the electrodes was varied from 2 to 5 mm. A titanium plate 5x5 mm was chosen as the irradiated target. The target does not have direct electrical contact with the target retainer in order to avoid the supply of high potential to the experimental table when high-voltage electrodes are short-circuited.

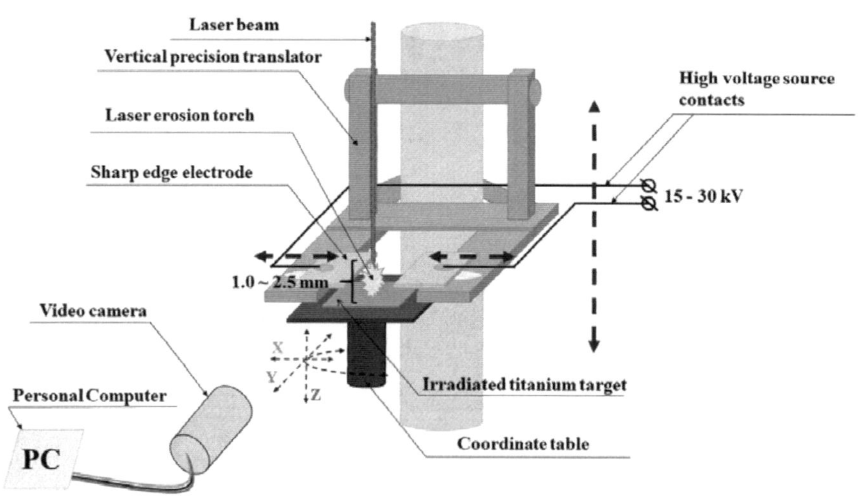

Figure 1. Experimental system for recording vertical deflections of ablated particles of a laser erosion torch.

A number of experiments were carried out in which the propagation velocity of the laser beam (10 mm/s and 2 mm/s), the laser pulse energy (75 mJ and 140 mJ), the distance of the high-voltage electrodes from the target surface, and the distance between the high-voltage electrodes were changed. Electrical breakdown between electrodes with a connected potential occurs during prolonged exposure as a result of the deposition of a significant amount of ablated titanium particles on the surface of the electrodes. The vector of motion of ablated particles in an electrostatic field after a certain period of time is directed towards the negative high-voltage electrode. With a decrease in the distance between the high-voltage electrodes and the target, an earlier shift of the vector of motion of the particles of the laser erosion plume towards the minus contact is observed.

3. Results and discussion

The photographs obtained allow us to make a conclusion about the effect of an electrostatic field with a strength of 15 kV/cm on the propagation profile of ablated particles of a laser erosion plume (Figure 2). The change in the trajectory of motion occurs smoothly - as the particle loses the imparted energy.

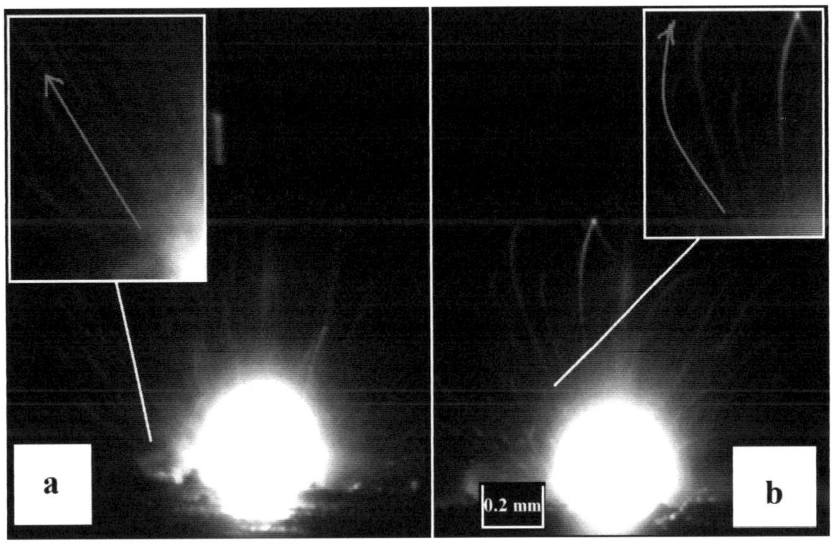

Figure 2. The process of propagation of ablated particles of a laser erosion plume; a - in the absence of a high-voltage potential on the electrodes, b - there is a high-voltage potential on the electrodes.

The obtained micrographs should be used to deduce the degree of influence of the electrostatic field on the ablated particles. Since the exposure time of a video camera frame is 16.65 ms when shooting 60 frames per second, this value was taken as the time of motion of the ablated particles in the working zone of the laser beam t_{pr} until a significant part of the energy imparted to them is lost. Next, we calculate the average height of motion of the ablated particles in the electrostatic field h_{ph} due to the correspondence of the actual dimensions of the laser processing area with its dimensions in the obtained micrographs.

Let us calculate the average speed of movement of ablated particles from the area of exposure to laser radiation using the following equation (1):

$$v_{pr} = \frac{h_{ph}}{t_{pr}} \quad (1)$$

Knowing the speed of motion of the ablated particles v_{pr}, it is possible to calculate the average time of motion of the ablated particles $t_{pr\,f}$, which, under certain conditions, will indicate the degree of influence of the electrostatic field on the front of propagation of the laser erosion plume. For this, it is necessary to calculate the height of free motion of the ablated particles h_{ph2} without taking into account the segment after the change in the direction of motion from the influence of the electrostatic field. Next, we calculate the average time of motion of the ablated particles before the effect of the electrostatic field on the vector of their motion $t_{pr\,f}$ according to the equation (2):

$$t_{pr\,f} = \frac{h_{ph2}}{v_{pr}} \quad (2)$$

All calculated data on the deviation of the front of propagation of particles of the laser erosion plume from the presence of an electrostatic field when changing the parameters of the laser beam and pulse energy are shown in Table 1:

Table 1. Results of calculating the deflection of ablated particles of a laser erosion plume under the action of an electrostatic field.

№	h_{ph} [mm]	v_{pr} [m/s]	h_{ph2} [mm]	$t_{pr\ f}$ [ms]
1	0.54	0.034	0.42	12.35
2	0.61	0.036	0.48	13,33
3	0.66	0.039	0.51	13,07
4	0.84	0.051	0.72	14,12
5	0.67	0.040	0.41	10,25
6	0.86	0.052	0.6	11.54

The final calculations make it possible to more accurately observe the effect of 15 kV/cm electrostatic field lines on the propagation front of the laser erosion plume above the titanium target surface when the laser beam repetition parameters and pulse energy change.

4. Conclusion

With an increase in the energy of the laser pulse and a decrease in the velocity of the laser beam, an increase in the front of propagation of the laser erosion plume by 0.3 mm was observed. With a decrease in the distance between the high-voltage electrodes and the target, an earlier displacement of the vector of motion of the particles of the laser erosion plume towards the negative contact is observed. The use of an electrostatic field made it possible to reduce the glow intensity of the laser erosion torch; also, a significant increase in the layer of deposited particles was observed on the electrode plates, which indicates the effective removal of ablation products from the region of propagation of the laser beam. Thus, the use of an electrostatic field makes it possible to increase the amount of energy delivered to the surface of the processed material by reducing the proportion of radiation scattered by suspended particles as a result of ablation of the processed material.

Acknowledgments

The study was carried out using the equipment of the interregional multispecialty and interdisciplinary centre for the collective usage of promising and competitive technologies in the areas of development and application in industry/mechanical engineering of domestic achievements in the field of nanotechnology (Agreement No. 075-15-2021-692 of August 5, 2021). This work was also partially supported by the framework of the state task of VlSU № 0635-2020-0013.

References

[1] Kononenko T V, Freitag C, Sovyk D N, Lukhter A B, Skvortsov K V and Konov V I 2018 *Opt. and Lasers in Eng.* **103** 65–70
[2] Vasiliev S V and Ivanov A Yu 2012 *Quant.electr.* **42(2)** 170–74
[3] Freitag C, Pauly L, Förster D J, Wiedenmann M, Weber R, Kononenko T V, Konov V I and Graf T 2018 *Adv.Optical Tech.* **7(3)** 157–63
[4] Lyubimov V V, Kozyr D V and Gnidina I V 2015 *Izvestiya Tula State Univer. Tech. sci.* **10** 82
[5] Chkalov R V, Khorkov K S, Kochuev D A, Zolotov A N and Prokoshev V G 2019 *J. Phys.: Conf. Ser.* **1164** 012009
[6] Chernikov A S, Kochuev D A, Voznesenskaya A A and Egorova A V 2021 *J. Phys.: Conf. Ser.* **1942** 012024
[7] Egorova A V, Kochuev D A, Chernikov A S, Mikhalevsky V A, Kireev A V and Khorkov K.S. 2021 *J. Phys.: Conf. Ser.* **2077** 012003
[8] Kochuev D. A, Chkalov R V and Vasilchenkova D G 2020 *Int.Conf. Laser Optics (ICLO)* 9285885
[9] Byeon J H, Hwang J, Park J H, Yoon K Y, Ko B J, Kang S H and Ji J H 2006 *J. of Aerosol Sci.*

37(11) 1618–28
[10] Huang S H and Chen C-C 2001 *Aerosol Sci. & Tech.* **35(4)** 792–804
[11] Wu S, Zou J, Xiao R and Li F 2015 *Laser Phys.Let.* **12(4)** 046006
[12] Ribic, B Palmer T A and DebRoy T 2009 *Int. Mat. Reviews.* **54(4)** 223–44
[13] Kochuev D A, Raznoschikov A S and Chkalov R V 2020 *IOP Conf. Ser.: Mater. Sci. Eng.* **969** 012034
[14] Chernikov A S, Kochuev D A, Voznesenskaya A A, Egorova A V and Khorkov K S 2021 *J. Phys.: Conf. Ser.* **2077** 012002
[15] Ivashchenko A, Tarasova M and Davidov N 2019 *Mat. Today: Proc.* **11** 465–80
[16] Kochuev D A 2020 *Bulletin of the Russian Academy of Sci.: Phys.* **84(3)** 330–35
[17] Voznesenskaya A A, Kireev A and Ivashchenko A 2020 *Solid State Phenomena. – Trans Tech Publ.Ltd* **299** 205–09
[18] Kochuev D A, Chernikov A S, Khorkov K S, Chkalov R V, Voznesenskaya A A, Prokoshev V.G 2019 J. *Phys.: Conf. Ser.* **1331** 012010
[19] Grigoriu C, Wang W N, Martin D and Biswas P 2012 *Aerosol and Air Quality Research.* **12(5)** 673–82

Journal of Physics: Conference Series **2316** (2022) 012012 doi:10.1088/1742-6596/2316/1/012012

Experimental determination of the spectral absorption coefficient of a plasma-erosion torch on the surface of carbon-bearing materials

A F Galkin, S V Zhirnova, V M Lipatov and E L Shamanskaya

Vladimir State University, 87 Gorky Street, Vladimir 600000, Russia

E-mail: svetlanna_vik@mail.ru

Abstract. In the work, the spectral coefficient of the plasma-erosion torch on the glassy carbon surface was determined on the basis of the Wien violation law and the experimental determination of the brightness temperature. The possibility of using a micropyrometer to determine the spectral factor can be a plasma-erosion torch, with the help of which (spectral absorption coefficient) it is rather quick to determine the thermodynamic temperature of the plasma-erosion torch.

1. Introduction

Currently, many different measuring transducers and devices are used to measure temperatures, which are based on different principles of operation. Micropyrometers are devices designed for non-contact measurement of body temperature. The principle of operation of this device consists in measuring the power of thermal radiation of an object, mainly in the ranges of infrared radiation and visible light. Pyrometers have a number of advantages over other devices that are used to change temperatures. Since the measurement takes place in a non-contact manner, therefore, many disturbing factors are eliminated, for example, interaction with an object. They are also easy to use, rather compact, have a relatively low cost and high reliability.

Temperature measurement in some works was carried out with an MP1001 micropyrometer [1-3]. The range of measured brightness temperatures is from 2000 °C to 5000 °C. Temperature measurements are carried out in two spectral intervals in the region of wavelengths of 0.9 and 0.65 μm.

The brightness temperature of a body in rays with a wavelength λ is such a temperature of an absolutely black body T_b, at which the brightness of both bodies in the light of the same wavelength is equal to each other.

The spectral absorption coefficient $\varepsilon_{\lambda,T}$ is an important parameter for determining the actual temperature of an object from the measured brightness temperature. The intensity of thermal radiation of real bodies depends on the physical nature of the body and the spectral absorption coefficient, the value of which is determined by the temperature and state of the surface [4]. The value of the absorption coefficient changes with changes in the state of the surface and temperature. Thermal radiation occurs at any body temperature and has a continuous spectrum, the position of the maximum of which depends on the temperature of the substance. For real bodies, the brightness temperature is always less than the real one [5].

The relationship between actual and brightness temperature is described:

$$\frac{1}{T_b} - \frac{1}{T} = \frac{\lambda}{C_2} ln \frac{1}{\varepsilon_{\lambda,T}} \quad (1)$$

Where λ – used radiation wavelength, μm; C_2 = 14388 μm·K – radiation constant; $\varepsilon_{\lambda,T}$ – spectral absorption coefficient of the body; T_b – is the brightness temperature; T – is the thermodynamic temperature.

The purpose of this work is to show the possibility of using a micropyrometer to determine the spectral absorption coefficient of a plasma-erosion plume.

2. Experimental installation

In this work, the spectral absorption coefficient of the plasma-erosion torch, formed under the action of repetitively pulsed laser radiation on the surface of glassy carbon, is determined. The diagram of the experimental installation is shown in Figure 1. In the figure, the numbers indicate: 1 – repetitively pulsed laser CLW50; 2 – glassy carbon sample; 3 – ASP-75 spectrometer for obtaining the emission spectrum of the plasma-erosion torch (MP-1001 micropyrometer for measuring the brightness temperature of the plasma-erosion torch).

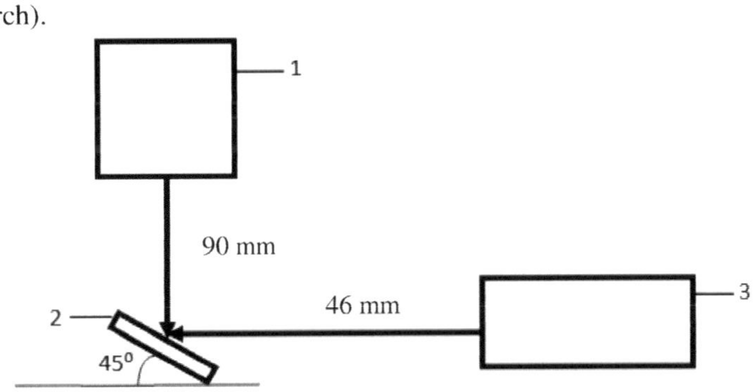

Figure 1. Schematic of the experimental installation

A repetitively pulsed CLW-50 laser was used as a power laser. Radiation of a Nd: YAG laser with an operating wavelength of 1064 nm, leaving the laser emitter via a fiber-optic cable, is fed into an optical-mechanical system to form a spot of the desired diameter in the affected area. An ASP-75 spectrometer was used to obtain the spectrum of the plasma erosion torch.

Figure 2 shows the appearance of the experimental setup. In the figure, the numbers indicate: 1 – ASP-75 spectrometer; 2 – a vacuum chamber containing a glassy carbon sample; 3 – pulse-periodic laser CLW50.

Figure 2. Appearance of the experimental installation

The pulse energy (E) varied from 1 J to 2 J, the pulse repetition rate (f) varied from 45 Hz to 50 Hz, and the pulse duration (τ) was 2 ms. The laser beam diameter was 0.4 mm. The brightness temperature was measured using an MP-1001 micropyrometer at a wavelength of 650 nm.

Figure 3 shows an image of a plasma-erosion torch of the surface of a glassy carbon target in air. The numbers indicate: 1 – glow near the target surface; 2 – glow at a distance of ~ 3 cm from the target surface; 3 – glassy carbon sample. Laser radiation parameters: $E = 2$ J; $f = 45$ Hz; $\tau = 2$ ms..

Figure 3. Spatial glow of an erosion laser torch on the surface of a glassy carbon target

3. Experiment Results

In the course of the experiment, several spectra of the plasma-erosion torch were obtained at various parameters of laser radiation acting on the glassy carbon surface.

As an example, Figure 4 shows the emission spectra of a plasma-erosion torch obtained with the following parameters of laser radiation: (a) $E = 2$ J, $f = 45$ Hz, $\tau = 2$ ms; (b) $E = 1$ J, $f = 50$ Hz, $\tau = 2$ ms. The thick lines in the figures show the spectrum of thermal radiation of the plasma-erosion torch.

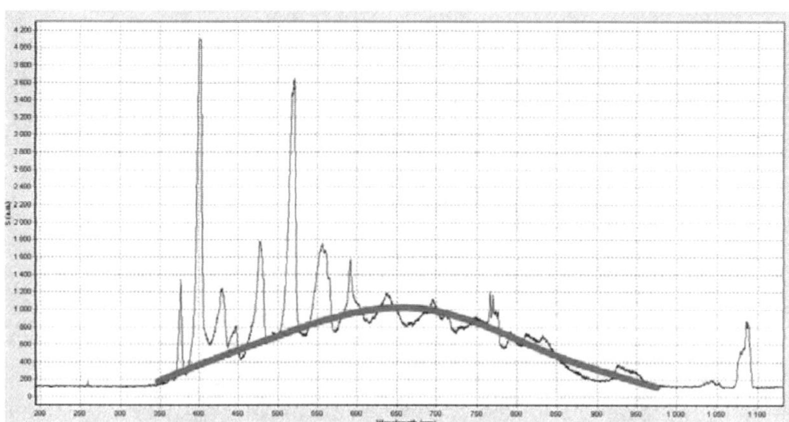

(a)

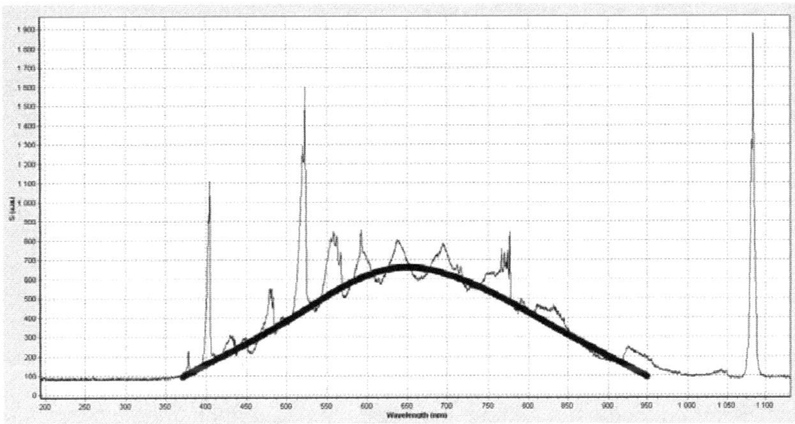

(b)

Figure 4. Radiation spectrum of a plasma-erosion torch

According to Wien's displacement law, the thermodynamic temperature was calculated on the basis of spectrograms: for the parameters of laser radiation $E = 2$ J, $f = 45$ Hz, $\tau = 2$ ms, the average temperature was 4480 K; for laser radiation parameters $E = 1$ J, $f = 50$ Hz, $\tau = 2$ ms – 4300 K.

The brightness temperature of the plasma-erosion torch, obtained using an MP-1001 micropyrometer, at the indicated parameters of laser radiation was 4300 K and 3500 K, respectively.

On the basis of the actual temperature obtained according to Wien's law and the measured brightness temperature according to formula (1), the spectral absorption coefficient of the plasma-erosion torch ε was calculated. The average values of the spectral absorption coefficient of the plasma-erosion plume, obtained in the calculation, were 0.88 for the laser radiation parameters $E = 2$ J, $f = 45$ Hz, $\tau = 2$ and 0.48 for the laser radiation parameter $E = 1$ J, $f = 50$ Hz, $\tau = 2$ ms, respectively.

4. Conclusion

In this work, the spectral absorption coefficient of the plasma-erosion torch near the glassy carbon surface was determined on the basis of Wien's displacement law and the experimental determination of the brightness temperature. Thus, it has been shown that a micropyrometer can be used to determine the spectral absorption coefficient of a plasma-erosion torch, with the help of which (spectral absorption coefficient) it is possible to fairly quickly determine the thermodynamic temperature of the plasma-erosion torch with acceptable accuracy.

Acknowledgments

This work supported by the RFBR project № 20-02-00515 and also partially supported by the framework of the state task of VlSU № 0635-2020-0013.

The study was carried out using the equipment of the interregional multispecialty and interdisciplinary center for the collective usage of promising and competitive technologies in the areas of development and application in industry/mechanical engineering of domestic achievements in the field of nanotechnology (Agreement No. 075-15-2021-692 of August 5, 2021).

References

[1] Arakelyan S, Galkin A, Zhirnova S and Osipov A 2015 *J. "Dyn. of Compl. Sys."* **9**(1) 48-50

[2] Arakelyan S, Zhirnova S, Galkin A, Kochuev D, Shamanskaya E and Khorkov K 2019 *EPJ Web of Conf* **220** 03034

[3] Arakelyan S, Zhirnova S, Galkin A, Kochuev D, Shamanskaya E and Khorkov K 2020 *Izvestiya RAS. Ph. Ser.* **84**(3) 436–38

[4] Potapov A 2016 *Sci. and peace* **1**(9) 10-22

[5] Timoshpolsky V I, Trusova I A, Kabishov S M, Klimovich G A and Ratnikov P E 2007 *Determination of the reduced emissivity and emissivity of a solid* (Minsk: BNTY Press) 18 p

Fractal models of the PbTe nanocluster structures on a solid surface

D N Bukharov, A S Abramov, O A Novikova, V D Samyshkin

Vladimir State University, Gorky Str. 87, 600000 Vladimir, Russia

E-mail:buharovdn@gmail.com

Abstract. A fractal approach is proposed for modeling systems of PbTe nanoclusters on a solid surface. The systems under consideration were generated by solid-phase laser modification. For the obtained samples, their fractal dimension was estimated. In accordance with the proximity to it, the corresponding chaotic fractal structures were selected.

1. Introduction

Nanocluster systems obtained on the basis of nanomodification of thin films of various nature have good prospects as a new element base for nanophotonics devices [1]. In particular, in the modern industry, semiconductor nanocluster lead telluride (PbTe) structures placed on a solid surface are basic materials for infrared optoelectronic systems and thermoelectric devices. This situation is associated with their unique electrical and optical properties, which are manifested in them at the nanoscale. So, for example, the transition to quantum size effects in such structures is observed already at sizes of 30-50 nm, while, for example, for silicon, for this purpose it is necessary to obtain structures with sizes of 4-5 nm. PbTe structures have a high thermoelectric coefficient and low thermal conductivity, therefore they are effective materials for the creation of industrial semiconductor thermoelements [2]. PbTe can be used to produce thin epitaxial films, the laser solid-phase nanomodification of which makes it possible to synthesize materials with topology-dependent properties. The laser synthesis method makes it possible to control the generated topology of the samples, and therefore to obtain the required properties [2].The use of the apparatus of mathematical modeling can be useful for preliminary assessment of the structure and topology of samples that can be synthesized in laser experiments.

2. Description of synthesis and experimental samples of nanocluster structures

The systems of semiconductor nanoclusters considered in this work were synthesized on the surface of an epitaxial PbTe film under the action of a continuous laser radiation of a YAG: Nd3 + laser with a wavelength of 1.06 μm, with a power of 5-10 W, while scanning a laser beam with a diameter of 30-50 μm at a speed of 60-80 μm / s over the sample surface [3]. The nanostructures induced in the laser experiment were studied using a Quanta 200 scanning electron microscope (SEM) and an Integra-Aura 3D probe nanolaboratory (AFM images). A detailed study of the obtained nanofilms, performed using tools in the MATLAB software environment, demonstrated their fractal nature. Calculation of fractal dimension, carried out in the approximation of boxcounting [4], made it possible to correlate islands of nanostructured thin-film cluster structures with known fractals with an accuracy of 0.03.

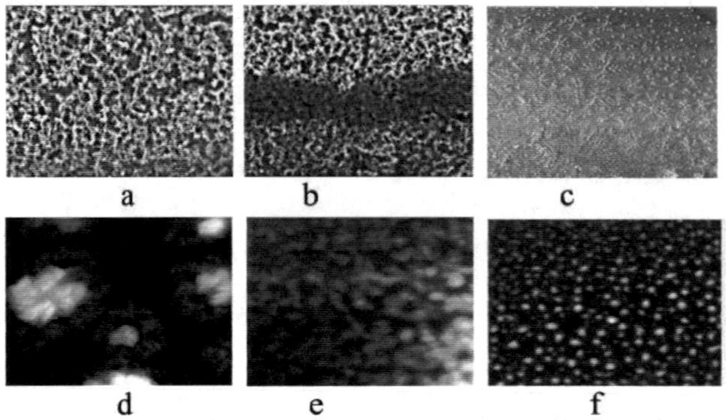

Figure 1. Images of the obtained fractal cluster structures: diffusion-limited aggregation (a), Brownian motion (b), snowflake (c), Koch Island (d), Julia set (e), percolation (g).

Table 1. Dimensions of samples and corresponding fractals

Sample type from Figure 1	Calculated dimension of the nanocluster structure (accurate to 0.01)	Corresponding fractal	Theoretical dimension of fractal boundaries (accurate to 0.01) [5,6]
a	1.87	DLA	1.07-2.02
b	1.53	Brownian motion	1.60
c	1.38	Star	1.40
d	1.72	Koch Island	1.63
e	1.24	Julia's set	1.20

Table 1 summarizes the calculated dimensions of the samples from Figure 1 and the corresponding values typical for standard fractals.

3. Modeling the structure of samples

The classification of the obtained samples in accordance with their fractal dimension makes it possible to determine the basic models that, in the first approximation, are applicable to describe the structure of the samples obtained by us.

In connection with the obvious fractal nature of the obtained nanoclusters, the approaches of fractal geometry are applicable for modeling their structure. Today there are a number of algorithms for generating fractal objects. The processes of formation of the systems of nanoclusters are rather complex and are caused by the influence of many factors that are difficult to take into account. Therefore, adequate models describing the structure of such samples should not be completely deterministic and contain random parameters that allow taking into account random factors. Such properties are possessed by models based on randomized fractals, which can be generated by various methods.

Stochastic geometric fractals, for example, based on the Koch snowflake, Star, Sierpinski triangle, Hilbert fractal [5], etc., as well as structures based on algebraic fractals - Julia and Mandelbrot sets [6] can be used as models of the first approximation to describe the topology of nanoclustes. The fractals listed above make it possible to simulate the shape of nanoclusters, but, unfortunately, it is rather difficult, and often even impossible, to introduce or explicitly take into account the physical parameters of the system and the conditions for obtaining films, for example, such as the temperature of the system.

On the other hand, there are algorithms for generating fractals that make it possible to take into account the physical processes that lead to the formation of nanoclusters rather obviously or explicitly. Analysis of the results of experiments on the preparation of nanoclusters made it possible to conclude that the main process influencing the shape of nanoclusters was thermal diffusion, in which nanoparticles moved, collided, aggregated and formed the film structure. Thus, consideration of such processes makes it possible to introduce the physical parameters of the system into the model.

Within the framework of the fractal approximation, structures synthesized due to thermal diffusion processes can be modeled on the basis of fractals of Brownian motion, diffusion-limited aggregation, percolation mechanism, as well as their modifications and variation of the initial and / or boundary conditions. The structure of films formed as a result of thermal diffusion processes at an early stage of formation, with a sufficient degree of agreement with experimental samples, can be described within the framework of fractal Brownian motion (FBD) [7], modifying the standard version of which one can take into account temperature changes in the system and the effect of laser action [8].

For well-formed nanocluster samples with a structure related to the dendritic one, the model of a diffusion-limited aggregation (DLA) fractal [9], implemented by the cellular automaton method [10], is applicable.

The percolation fractal model [11] was used to describe nanoclusters with a percolation nature of the surface relief. Fractal nanofilms with a percolation structure and good filling of nanoclusters can be modeled in the directed percolation approximation [ ]. This technique allows generating Eden clusters [12].

Models of nanoclusters were developed on the basis of the fractals indicated in Table 1. Models of nanoclusters were developed on the basis of the fractals indicated in Table 1. Some of the model structures are shown in Figures below.

Figure 2 shows the Koch island models generated by randomizing a regular fractal (Figure 2a). To obtain a stochastic fractal with, the rotation angle was randomized [13]. Figures 2b and 2c show Koch island after 4 and 5 iterations, respectively.

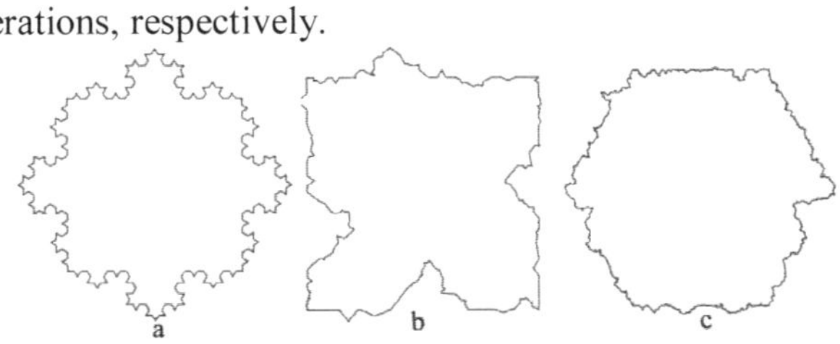

Figure 2. Koch fractal model: regular fractal after 5 iterations (a), randomized after 4 iterations (b), randomized after 5 iterations (c).

Figure 4 shows model images of nanocluster systems based on the Harter-Hethway Dragon fractals (Figure 4a - regular fractal, Figure 4b - randomized) and Sierpinski triangle (Figure 4c - regular fractal, Figure 4d - randomized).

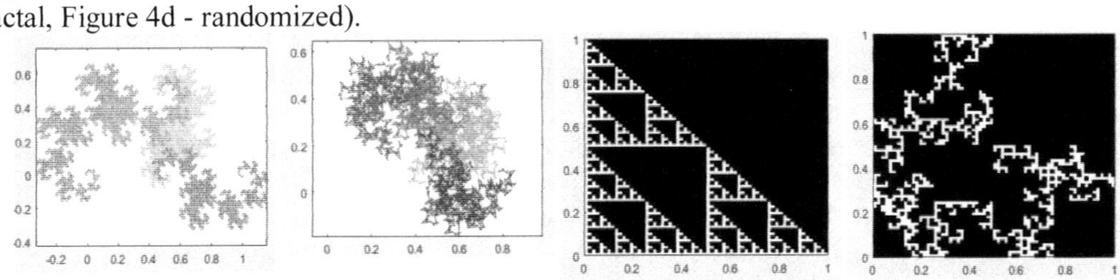

a b c d

Figure 3. Geometric fractal model: Harter-Hethway Dragon regular fractal after 12 iterations (a), Harter-Hethway Dragon randomized fractal after 12 iterations (b), Sierpinski triangle regular fractal after 7 iterations(c), Sierpinski triangle randomized fractal after 7 iterations (d).

The variety of forms of models based on the Julia set, based on the relation $z = z^2 + c$ for different values of the control parameter c, can be traced in Figure 4. The shape of fractal structures varies from dendritic to rounded and from single to multiple.

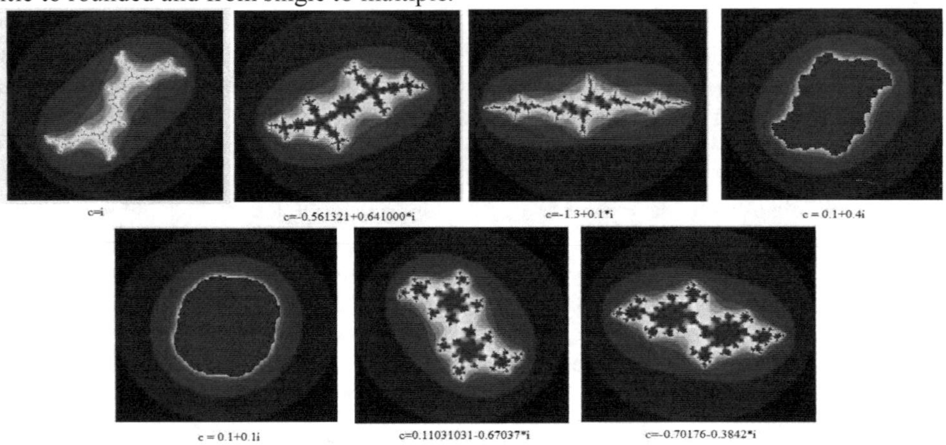

Figure 4. Geometric fractal model: Harter-Hethway Dragon regular fractal after 12 iterations (a), Harter-Hethway Dragon randomized fractal after 12 iterations (b), Sierpinski triangle regular fractal after 7 iterations(c), Sierpinski triangle randomized fractal after 7 iterations (d).

The figures below show models of nanoclusters based on fractal structures obtained in the diffusion approximation. Figure 5 shows a model of nanoclusters in the approximation of fractal Brownian motion at a medium temperature of 293 K (Figure 5a) and 373 K (Figure 5b). This calculation indicates that more compact cluster aggregates are formed at low temperatures.

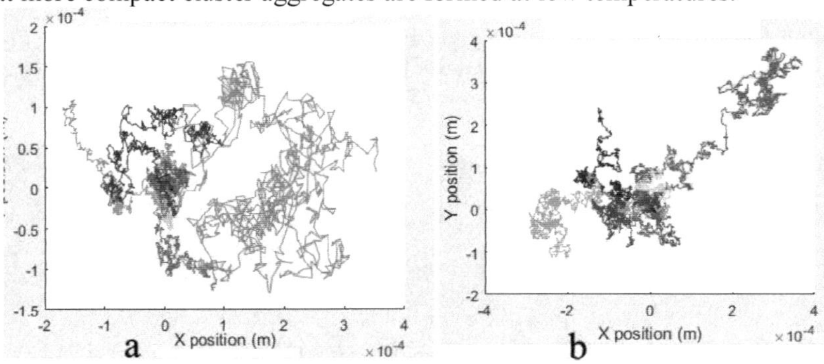

Figure 5. FBD model: temperature 293 K (a), 373 K (b).

Figure 6 shows the DLA model of nanoclusters at a constant value (Figure 6a) and varying in time (Figure 6b) towards a decrease in the main control parameter of the model - the probability of particle adhesion to an already formed structure. Figure 6 shows the DLA model of nanoclusters with a constant value (Figure 6a) and time variation (Figure 6b) in the direction of decreasing and increasing (Figure 6d) of the main control parameter of the model - the probability of particle adhesion to an already formed structure. In Figure 6a, the sticking probability value throughout the entire calculation remained constant, equal to 0.9. In Figure 6b, its value decreased steadily from 1 to 0.01. Figure 6e of obtaining at its uniform increase from 0.01 to 1. For this case, a system of particles randomly located in the computational domain was chosen as the initial aggregate (Figure 6c).

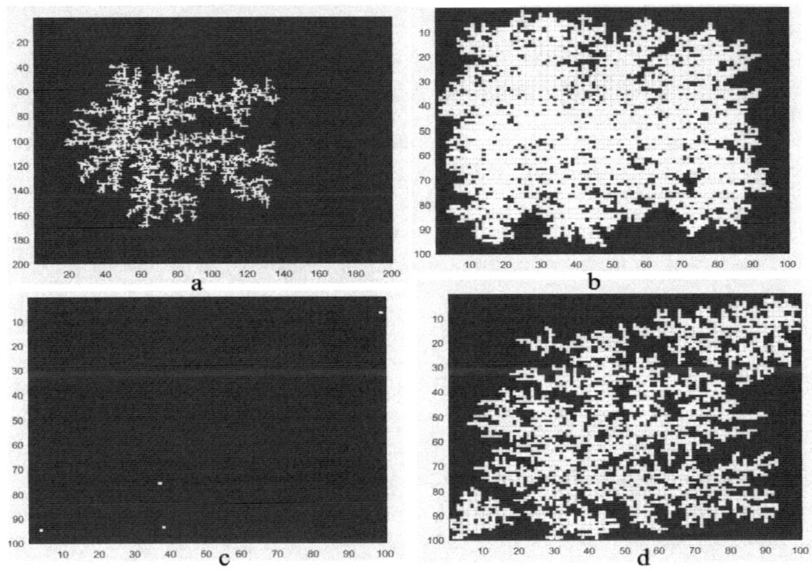

Figure 6. DLA model: constant probability of particle adhesion 0.9 (a), decreased variable probability of particle adhesion (b), initial aggregate(c), increased variable probability of particle adhesion.

Figure 7 shows the simulation results in the percolation approximation: the lattice percolation model (Figure 7a) and the directional percolation model from a point source (Figure 7b). The main parameter of the model (p) is the probability of transferring the attribute of the neighboring cell of the computational domain. The lattice percolation model makes it possible to model the internal structure, and the directed percolation approach makes it possible to generate fractal boundaries and take into account the shape and number of starting sources, starting from which the aggregate is formed.

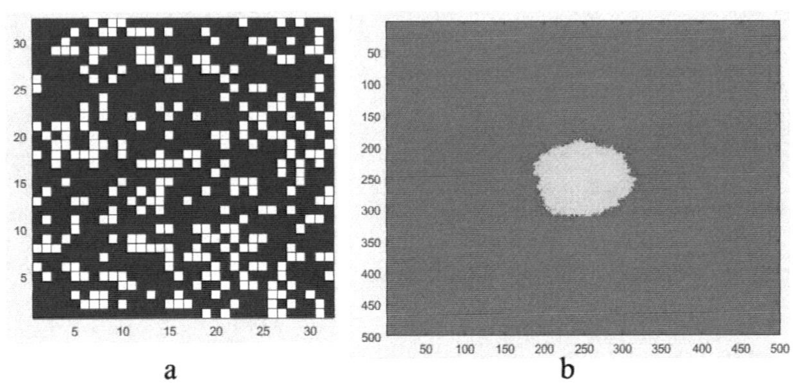

Figure 7. Percolation model: lattice percolation p=0.735 (a), directional percolation p=0.01(b).

4. Conclusion

The proposed models are of a stochastic nature, therefore, qualitative assessment methods are applicable to assess their accuracy. Nevertheless, it is possible to quantify and correlate the fractal dimensions of the generated figures and images of samples from experiments. This relationship is important because there is a dependence of the properties of nanocluster structures (for example, electrically conductive or optical) and their fractal dimensions. For the generated figures, their fractal dimension was estimated by the boxcounting method and the ratio of its value to the dimensions of the samples. This comparison showed a satisfactory degree of agreement; therefore, the proposed approach can be useful for evaluating the experimental production of nanostructures, in particular, PbTe nanoclusters.

Acknowledgments
The study was carried out using the equipment of the interregional multispecialty and interdisciplinary center for the collective usage of promising and competitive technologies in the areas of development and application in industry/mechanical engineering of domestic achievements in the field of nanotechnology (Agreement No. 075-15-2021-692 of August 5, 2021).

This work was also partially supported by the framework of the state task of VlSU № 0635-2020-0013.

References

[1] F Monticone, A Alù 2017 *Rep. Prog. Phys.* **80** 036401.
[2] S P Zimin, E S Gorlachev 2011 Nanostructured Lead Chalcogenides: Monography (Yaroslavl: YarSU).
[3] SM Arakelyan et al 2015 Introduction to Femtonanophotonics. Fundamentals and laser methods for controlled production and diagnostics of nanostructured materials: study guide (Moscow: Logos).
[4] G Gonzato 1998 *Computers & Geosciences* 24.
[5] K J Falconer 1990 Fractal geometry: Mathematical foundations and applications (Chichester: Wiley).
[6] J Feder 1988 Fractals (New York, London : Plenum Press).
[7] G Slade 2019 *Proc.R. Soc. A.* **475** 2221.
[8] D N Bukharov et al 2019 Mathematical and computer modeling: collection of materials of the VII International scientific conference dedicated to the memory of S.S. Efimova (Omsk: Publishing house Om. state University).
[9] J Mroczka, M Woźniak, F R A Onofri 2012 *Metrol. Meas. Syst.* **XIX** 3.
[10] D A Zaitsev 2017 *Theoretical Computer Science* **666** .
[11] D I Iudin, E V Koposov 2012 Fractals: from simple to complex(N. Novgorod: NNGASU).
[12] E Agyingi, L Wakabayashi, T Wiandt, S Maggelaki 2018 *Processes*. **6**.
[13] V A Saroka, K G Batrakov and L A Chernozatonskii 2014 Physics of the Solid State **56** 10.

Structure and oxygen-sensitive properties of thin films of natural carbon

A A Kovalchuk

Laboratory of Physical and Chemical Nanomaterials Study, Institute of Geology Karelian Research Centre RAS, Petrozavodsk, 185910, Russian Federation

E-mail: kovalchuk_aa@krc.karelia.ru

Abstract. Thin film carbon nanomaterials are widely studied and used as functional, conductive, reflective, and sensor coatings. In this work, films of natural graphene-based carbon are studied as a sensitive material for a gas sensor operating close to room temperature. The structural features of the films were studied by scanning electron microscopy and Raman spectroscopy. It has been found that the deposition of natural carbon on substrates with a conductive coatings of indium oxide and of indium tin oxide allows to obtain a thin films, represented by homogeneous carbon nets, in the nodes of which are globular nano-sized particles. Microwave conductivity measurements were carried out using a $\lambda/4$ coaxial resonator based on a symmetric two-wire line. The results demonstrated that films on substrates with deposited indium oxide and indium tin oxide desorb oxygen from various structural levels of graphene-based carbon.

1. Introduction

Thin-film carbon materials have proven themselves well as elements of semiconductor devices, gas sensors and coatings to enhance surface functionality. Nanocarbon materials such as carbon nanotubes [1 – 3], porous carbon [4] and graphene [5, 6] often act as the sensitive materials in solid state gas sensors. The emergence of a new generation of gas sensors is largely due to the exceptional sensitivity of nanocarbon materials, which allows the detection of toxic gases in concentrations of up to 1 ppb.

One of the main directions in the manufacture of gas sensors is the search for methods to increase the specific surface area of a sensitive material. One of the options for increasing the specific surface area of a sensitive material is the formation of a nanostructured gas-sensitive layer (thin film) on a substrate. In this paper, films of graphene-like carbon were studied as a sensitive material and proposed for a gas sensor.

2. Materials and method

Shungite carbon is a natural carbon allotrope with a multilevel structure. Its structure is formed by sequential aggregation of graphene fragments ~ 1 nm in size.

Thin films were obtained from shungite nanocarbon by sublimation [7] of shungite powder in a heat chamber. The process of obtaining films is described in more detail in [8].

Carbon powder made from type I shungite from the Shunga deposit (Karelia, Russia) [9] used as the source was prepared by the technology described in [10]. Particle sizes of the powder were 0.01–1.0 μm.
The quality of the films was controlled by scanning electron microscopy and Raman spectroscopy. Raman spectra were recorded with laser radiation with the wavelength 532 nm and with a Raman shift in the range of 100-3500 cm^{-1}.
Microwave conductivity measurements were carried out using a $\lambda/4$ coaxial resonator based on a symmetrical two-wire line inside a circular air-insulated shield. The frequency range in which the measurements were carried out was 0.65 - 1.2 GHz. The measurements were carried out with continuous heating of the sample in the temperature range 290–360 K with an average rate of temperature change of 10 K/min. The temperature value was determined by analyzing the microwave absorption signal. During successive measurements, the samples of carbon films on substrates were cooled within 5 - 10 minutes to room temperature.

3. Results and discussion

3.1. SEM and Raman spectra

SEM data showed that films on a substrate with an In_2O_3-coating are represented by globular carbon nanoparticles forming three-dimensional (3D) networks in the space between In_2O_3 grains. The 3D network is located over the entire area of the carbon film and binds together large particles (0.5 - 1 μm) In_2O_3. Carbon nanoparticles were found on the In_2O_3 grains, collected in a two-dimensional (2D) network, as shown in the inset to Figure 1. The nanoparticle size was 50-100 nm.
Films on an ITO-coated substrate are characterized by larger particles at nodes of the network (450–1000 nm), as shown in the inset to Figure 1. The amount of NaCl on the surface of the film increased by an order of magnitude in comparison with the film on the In_2O_3-coating. On NaCl crystals, as on In_2O_3 grains, nanocarbon particles are observed in carbon films, forming 2D network.

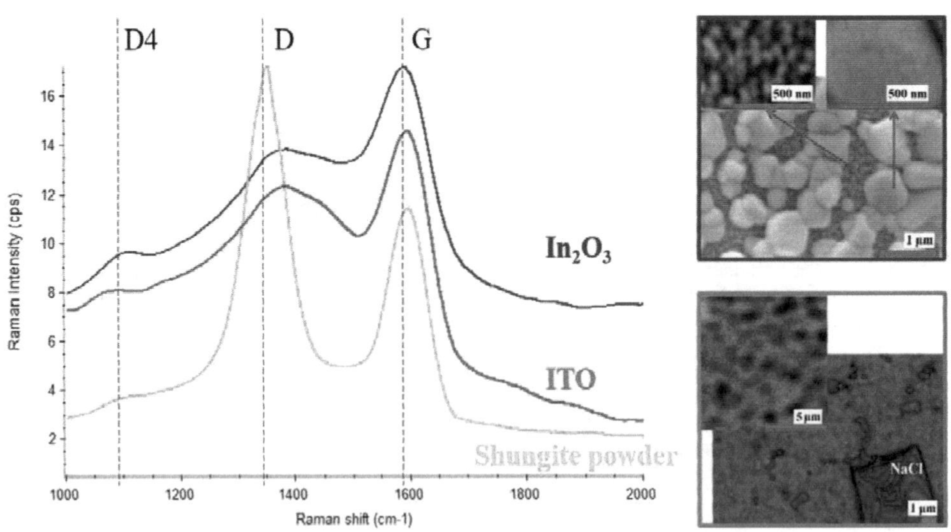

Figure 1. Raman spectra and SEM images of shungite powder, shungite carbon nanoparticles in films on the In_2O_3-coating and on the ITO-coating.

Figure 1 shows the Raman spectra for the studied thin films in comparison with the spectrum for shungite powder.

The Raman spectra of carbon films on substrates coated with In_2O_3 and ITO contain the main peaks (D and G bands) in the first order region, which are characteristic of noncrystalline carbon materials. By decomposing the peaks in the Omnic software (Thermo Scientific), the following positions of D and G peaks were obtained: 1391 and 1587 cm^{-1} for the film on the In_2O_3-coating, 1379 and 1589 cm^{-1} for the film on the ITO-coating, respectively. In the Raman spectra of films on In_2O_3- and ITO-coatings, a D4 band (~1100 cm^{-1}) appears, which is assigned to the deformation (bending) of graphene planes [11].
In both the spectra of the films an inversion of the main peaks of the D and G vibrations is observed in comparison with those for the initial powder. The intensity ratio of bands D and G decreased by several times, from which it can be concluded that the crystallites in the film are small and that there are defects.

3.2. Microwave absorption

The specific surface area of natural shungites in the initial state is relatively small [12], therefore, their use as sorbents requires of specific surface area development and modification.
The results of the modification of shungite in several ways are described in [13]. Thermal-oxidative activation, thermal activation in a reducing atmosphere, low-temperature modification by ozone, modification in oxygen and argon plasma of a glow discharge are efficient in surface modification. Thermal oxidative treatment leads to a high degree of carbonization of shungite carbon and an increase in the specific surface area of shungite carbon by almost two orders of magnitude. The ozonation results in an increase in adsorption capacity by several times. Treatment in argon plasma makes it possible to almost completely cleanse the surface of shungite carbon from functional groups, giving it hydrophobic properties. The disadvantage of these methods is their technological complexity and energy consumption.
Thin films of graphene-based carbon studied in this work were obtained under technologically gentle conditions: without using high temperatures (up to 1023 K) and catalysts [8]. The use of mild conditions for the preparation of carbon nanostructures is also evidenced by authors [14] who abandoned the arc discharge method in favor of sublimation in vacuum. They obtained carbon structures in the form of honeycombs with sp^2-bonds with high absorption capacity from graphite.
To study microwave conductivity in the temperature range above room temperature, the temperature dependences of the signal power P taken from the probe were measured when the base plane of the sample is parallel to the electric field.
The temperature dependences of carbon films on glass substrates with In_2O_3 and ITO are characterized by two distinct maxima in the range 300–325 K, as shown in Figure 2.

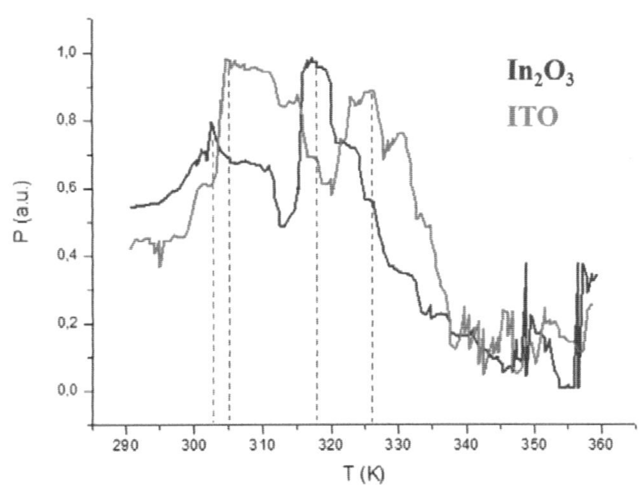

Figure 2. Temperature dependences of the reflected microwave signal intensity for carbon films on the In_2O_3-coating and on the ITO-coating.

The presence of temperature peaks in Figure 2 indicates a sharp increase in the resistance of the films under study on In_2O_3 and ITO coatings. In contrast to the results presented in [15], the temperature dependences of the studied carbon films on substrates coated with In_2O_3 and ITO are characterized by two distinct maxima in the range 315–335 K, which indicates the release of oxygen from various structural formations.

The granulometric composition of shungite powders affects the indices of their sorption activity [16]. The total porosity and specific volume of macropores increase with decreasing of grain size, which means free access to sorption centers located in the meso- and micropores of shungite. The size of carbon nanoparticles (see SEM images to Figure 1) of the film on the ITO coating is much larger than of the film on In_2O_3. This was reflected in the desorption curves: the finer the particles, the lower the desorption temperature.

Presumably, in the case of carbon films on substrates with In_2O_3 and ITO, the peaks in the temperature dependences (Figure 2) refer to oxygen emissions from various structural formations. The low-temperature peak (303 and 305K) is due to the desorption of oxygen located at contacts between globular particles in the film. The second peak (315 and 326 K) is connected with the release of oxygen, which is located in the distorted graphene planes of globular particles.

4. Conclusion

Thin films of graphene-based carbon were obtained under technologically gentle conditions: at temperatures below 1023 K and without catalysts.

It was revealed by SEM that the films deposited on substrates with conducting In_2O_3 and ITO contacts are formed of the network of shungite carbon nanoparticles. 2D network with carbon nanoparticles in the nodes appears on the faces of micron-sized crystal.

2D carbon network can be distinguished by dissolving the crystals, in the case of NaCl crystals.

The graphene-based structure of shungite carbon was detected by Raman spectroscopy in the films produced by the developed method.

The effect of oxygen desorption in the microwave wavelength range for films on substrates with deposited In_2O_3 and ITO was observed in the temperature range of 290–360 K. It was associated with the desorption of oxygen from the contacts between globular particles in the film, and with the release of oxygen from the distorted graphene planes.

A correlation was established between the size of nanoparticles in films on In_2O_3 and ITO – coatings and the temperature of oxygen desorption.

It is shown that replacing the conducting In_2O_3 substrate with ITO allows one to vary the characteristic temperatures of oxygen desorption.

Acknowledgments

This study was supported by the financial support of the Russian Foundation for Basic Research within the framework of the scientific project No. 20-53-04013/20 and state order project № AAAAA18118020690131-4.

References

[1] Kong J., Franklin N R., Zhou C., Chapline M G, Peng S, Cho K, Dai H 2000 Nanotube molecular wires as chemical sensors *Science* **287** 622–625.

[2] Collins P G, Bradley K, Ishigami M, Zettl A 2000 Extreme oxygen sensitivity of electronic properties of carbon nanotubes *Science* **287** 1801–1804.

[3] Rajavel K, Lalitha M, Radhakrishnan J K, Senthilkumar L, Thangavelu R, Kumar R 2015 Multiwalled carbon nanotube oxygen sensor: enhanced oxygen sensitivity at room temperature and mechanism of sensing *ACS App. Mater. Interfaces* **7** 23857–23865.

[4] Afify A S, Ahmad S, Khushnood R A, Jagdale P, Tulliani J M 2017 Elaboration and characterization of novel humidity sensor based on micro-carbonized bamboo particles *Sens. Actuators, B Chem.* **239** 1251–1256.

[5] Mendes R, Wrobel P, Bachmatiuk A, Sun J, Gemming T, Liu Z, Rümmeli M 2018 Carbon Nanostructures as a Multi-Functional Platform for Sensing Applications *Chemosensors* **6** 1–28.

[6] Varghese S S, Lonkar S, Singh K K, Swaminathan S, Abdala A. 2015 Recent advances in graphene based gas sensors *Sens. Actuators, B Chem.* **218** 160–183.

[7] Konkov O I, Mikhaylina A A, Prikhodko A V, Rozhkova N N 2016 Nanostructured membranes based on a natural carbon material *Journal of Optical Technology* **83** 286–289.

[8] Rozhkova N N, Kovalchuk A A, Konkov O I, Prikhodko A V 2018 Method of obtaining carbon film, Russian Federation Patent 2652206.

[9] Chou N H, Pierce N, Lei Y, Perea-López N, Fujisawa K, Subramanian S, Robinson J A, Chen G, Omichi K, Rozhkov S S, Rozhkova N N, Terrones M, Harutyunyan A R 2018 Carbon-rich shungite as a natural resource for efficient Li-ion battery electrodes *Carbon* **130** 105−111.

[10] Rozhkova N N 2011 Nanocarbon of schungites *Karelian Research Center of the RAS* Petrozavodsk 100.

[11] Ferrari A C, Robertson J 2004 Raman spectroscopy of amorphous, nanostructured, diamond-like carbon, and nanodiamond *Phil. Trans. R. Soc. Lond. A.* **362** 2477–2512.

[12] Filippov M M 2002 Shungite-bearing rocks of the Onega structure *Karelian Research Center of the RAS* Petrozavodsk 287.

[13] Mosin O V, Ignatov I I 2012 Composition and structural properties of natural fullerene-containing mineral shungite mined in Russia *Nanoengineering* **6** 17–23.

[14] Krainyukova E N, Zubarev E N 2016 Carbon honeycomb high capacity storage for gaseous and liquid species *Phys. Rev. Lett.* **116** 055501.

[15] Masterov V F, Prikhodko A V, Nemchuk N I, Shaklanov A A, Konkov O I 1997 Effect of oxygen on the electrophysical properties of some allotropic forms of carbon *Solid St. Phys.* **9** 1703–1705.

[16] Golub S L, Ulyanov A V, Buryak A K, Lugovskaya I G, Anufrieva S I, Dubinchuk V T 2006 Composition and sorption properties of shungite material *Sorption and chromatographic processes* **6** 748–763.

Topological photonics – 4D-laser technology for nanoelectronics on new physical principles to create the element base of next generation

A Kucherik[1], S Kutrovskaya[2], D Bukharov[1] and T Khudaiberganov[1]

[1]Vladimir State University, Gorky Str. 87, 600000 Vladimir, Russia
[2]International Center for Polaritonics, School of Science Westlake University, No.18 Shilongshan Road Cloud Town, Xihu District 310024 Hangzhou, Zhejiang PR China

E-mail: kucherik@vlsu.ru

Abstract. In the paper we consider femtonanophotonics of the topological controlled low-dimensional dynamic structures in thin films induced by laser radiation on the surface of solids, using the achievements of quantum technologies and nonlinear dynamics for different regimes of electroconductivity.
We completed several laser procedures for obtaining nanostructures and thin films with controllable topology. They occur under the development of different nonlinear processes in the system (thermodiffusion, gas-dynamic evaporation in pore-like structures with bubbles, ablation products, ballistic movement of the particles in liquid). A simple 2-steps mechanism for enhancement of quantum behavior (e.g. in electroconductivity) exists for different conditions.
First, when inelastic length $l_{inelastic} > a_{cluster}$ (size of cluster) we have no incoherent electron-phonon (e-ph) scattering, and coherent process occurs with scale ℓcoh.
Second, when de Broglie length $\lambda_{dB} \equiv \ell_{coh} > \Lambda$ (Λ – spatial period of the nanoparticle distribution) the coherent tunneling without loss occurs, and a long-range order with interference of the states takes place in the medium due to lattice structure in system.
Under such conditions, the electroconductivity enhancement can result in paradoxical phenomenon when electroconductivity in granulated structure may be higher than in monolith sample due to many surface/boundary units – like topological states in topological insulator.

1. Introduction

In the paper we consider femtonanophotonics of the topological controlled low-dimensional dynamic structures in thin films induced by laser radiation on the surface of solids, using the achievements of quantum technologies and nonlinear dynamics for different regimes of electroconductivity.

Currently, in practical aspect, all modern microelectronics is faced with problems of both technical and fundamental nature for technoprocesses.

In the first case, we are talking about the scaling limits, when with an increase in the density of the transistor arrangement, the heat release in the nanometer process increases and uncertainties arise due to quantum processes associated, for example, with tunnel currents [1]. In addition, the high cost of processes with a size of units of nanometers (see [2, 3]) and the requirements for the stability and reliability of their operation modes in the real conditions of the moving board location (vibrations, temperature variation and jumps, aggressive environment, oxidation, radiation etc.) determine the required practical scale of tens or even hundreds of nanometers (see, for example, [4]).

In the second case, in the fundamental aspect, the key drawback of the traditional von Neumann architecture is the fact that the commands for processing data and the data itself are stored together in memory, but the data processing operations and the execution of the actual calculations have a high speed in comparison with conversion to the memory [5]. In addition, there are natural fundamental limitations when the size of the designed element cannot be less than the lattice scale of the material (for silicon, for example, it is 0.5 nm).

An intermediate solution to these problems may be the transition to a neuromorphic architecture of a photon processor, when information is stored and processed simultaneously using photons of light [6]. This new hardware level is completely integrated into existing electronic circuits and functionally we are talking about hybrid systems «electrophysics+optics» [7].

A more global solution to the problem is the use of quantum technologies, including quantum infocommunication systems and quantum cryptography based on the achievements of quantum information advantages, and as the ultimate goal in future – the development of a quantum computer [8].

Taking into account the creation of an appropriate element base in nanoelectronic field by the achievements of modern laser experiments – femtonanophotonics [9], it is possible to control the spatio-temporal characteristics of such elements in the required direction to solve the problem of changing their functional properties in necessary directions. These, in fact, the 4D-technologies are the tool that should allow you to develop elements of logical systems on new physical principles. There are a number of directions for implementing these plans (see, for example, [10]).

In presented paper, we discuss one of the possible implementations along this problem – using nanophotonics approaches of low-dimensional topological thin-film structures, being, probably, a way for "Digital sovereignty" of Russia by unique kind hybrid systems and devices (electrophysics+optics).

Indeed, as demonstrated e.g. in our work [11], the electrical resistivity R of nanostructured samples can vary depending on the topology of structures deposited by laser methods on a solid substrate (for example, gold-Au nanoclusters) by more than 4 orders of magnitude when the nanocluster sizes change from 50 nm to 5 nm (Figure 1). In this case, everything is determined by the fractal dimension D of the obtained nanocluster structures (respectively, from, $D = 1.39$ to $D = 1.93$ by our calculation). Moreover, fairly uniform histograms of the size distribution of nanoclusters can be obtained quite easily, for example, at a scale of 10 nm±0.25 nm (see Figure 1 on the right).

Naturally, for such different topological structures, the optical transmission spectra of the substrates also change, and the Au-plasmon resonances shift/change in the region of 580 nm – Figure 2 (for monolithic Au-samples, the absorption wavelength of the plasmon resonance is 520 nm). At the same time, the maximum optical transmission is observed for $D = 1.93$ But the problems arise for such a granular structure (vs form, size and distribution, manylayers vs thickness) that not so easy to have one-way correspondence between topology and optical spectra – like in photonic crystals with varied parameters.

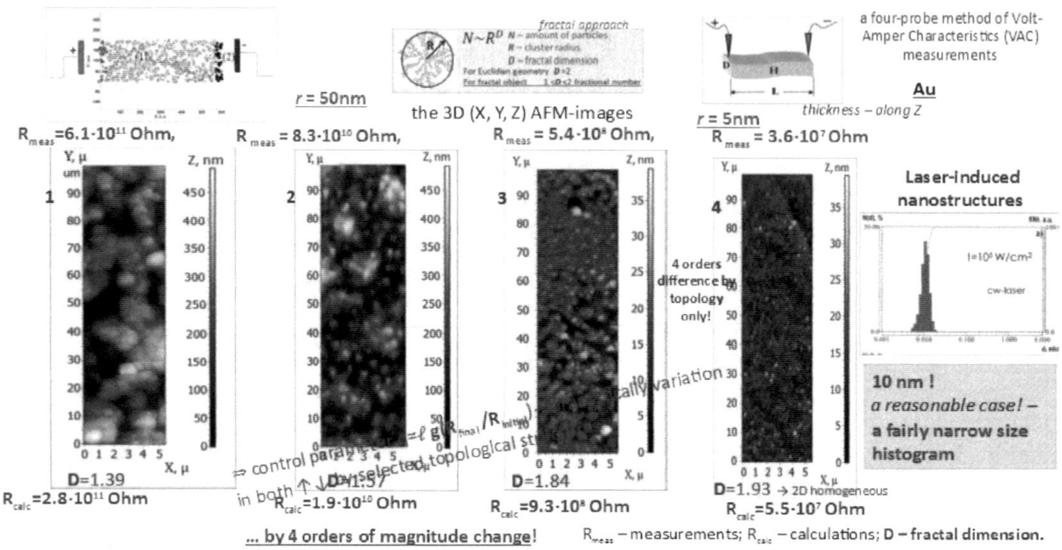

Figure. 1. Our measurements: electroresistance R_{meas} vs nanoparticle size (r) and topology (D) – over a wide range of the values. Here: R_{calc} – calculation result by adjustable fractal dimension. The narrow histogram over the size distribution of the nanoparticles is shown on the right.

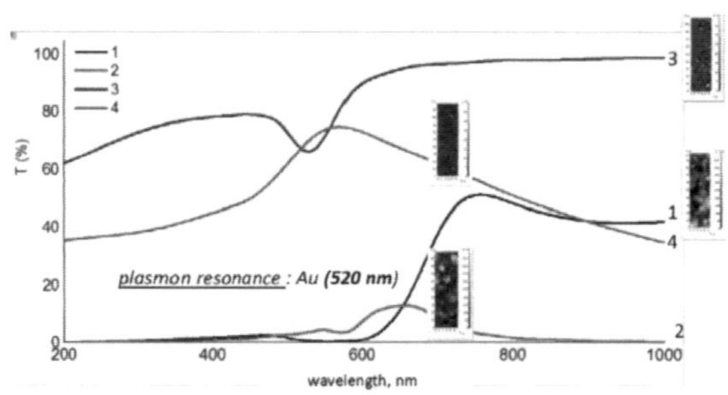

Figure. 2. The nanocluster structure should be correlated with the transmission spectra (vs structure: more conductivity in shell like system – wide optical spectra). Key parameters: sharpness (width and depth), and therefore a shift of spectra light propagation like it occurs through photonic structure.

The material of the paper is organized by following way. In next division 2 we discuss the 4D-Laser technology fabrication of new nanostructures and materials/cluster materials on solid surface, and consider the physical picture for several functional properties. In division 3 we analyze dendrite/fractal structures. In conclusion, we enumerate the summary collection over our different experiments on the topology peculiarities of the granulated metallic film deposited on dielectric substrates in frame of the topology photonics, and discuss as well the several perspective directions in the field.

2. The 4D-Laser technology fabrication of new nanostructures and materials/cluster materials on solid surface. Physical picture for several functional properties

Nanostructures and thin films with controllable cluster topology vs time depend on the laser pulses duration. In general, the interaction effects of solid targets with laser pulses of different durations for obtaining various nanocluster structures can be viewed as the possibility of synthesizing the 4D-objects, when the result depends not only on the stationary topological/geometric parameters of the system, but also on the dynamic interactions in the system leading to different final stable structures.

This is due to the fact that for different durations of laser pulses the specific mechanisms of nanostructuring are activated (i.e. Thermodiffusion, Gas-dynamic evaporation in pore-like structures with bubbles, Ablation products, Ballistic movement of the particles in liquid). Therefore, time plays the role of a control parameter responsible for phase transitions, as well as the spatial parameters do when nanoparticles (NPs) of various dimensions arise – from quantum dots (0D) to 3D nanostructures. In addition, for short laser pulses we have non-equilibrium/transient phase transitions over the steady-state pressure-temperature (PT) phase-diagram according to the laser trajectory of heating. Although the conventionality of this consideration is obvious (the equilibrium phase diagram cannot be used for non-stationary processes), but it allows to discuss the current trends and clarifying the basic physical picture.

Necessary conditions for quantum correlations are under expression for de Broglie lengths $\lambda_{dB}= h(mK_BT)^{-1/3} >> R\sim (V/N)^{1/3}$, where K_B – Boltzman constant, m – charged particle mass, T – temperature, h – Plank constant, R – an average distance/ interaction spatial scale between NPs, V – volume, N – number of NPs. For typical monolith semiconductors relation between effective electronic mass m_{eff} in solid and free electon mass m_0 is: $m_{eff}/m_0\sim0.1$, $K_BT=0.025eV$ (for the room temperature), and $\lambda_{dB}=25nm$ (in contrast with metal, where $\lambda_{dB} < 1nm$). But for certain materials this value of m_{eff}/m_0 is less in 10 times, e.g. for InSb $m_{eff}=0.013\ m_0$ ($\sim10^{-32}kg$) – see e.g. [4]. Therefore, we can talk about macroscopic quantum objects, especially for granulated structures when a spatial dispersion plays a principal role: $m_{eff} = \hbar^2 * \left[\frac{d^2\varepsilon}{dk^2}\right]^{-1}$, ε (k) – dispersive curve for the structure.

The formation of topological nanocluster structures in multilayer thin films on a solid surface in a controlled manner was realized in our experiment by laser methods according to a two-stage scheme (see, for example, [11]) – Figure 3.

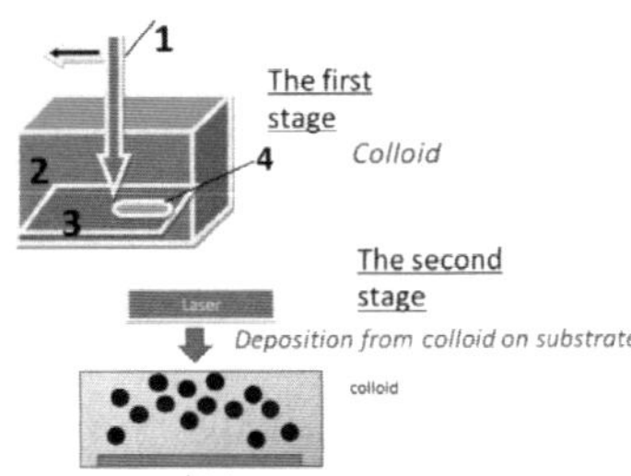

Figure. 3. Two-stage scheme for obtaining nanostructures on the surface of a solid surface (the explanation is given in the text).

First, the synthesis of nanoparticles/nanoclusters was carried out during laser ablation (1) of a target (3) from different materials placed in certain liquids (2) to obtain a colloidal system. In this case, laser radiation with different pulse durations was used – with a repetition rate of 20-100 kHz from msec to fsec – with an average power of 10^6-10^7 W/cm^2.

Second, the deposition of nanostructures from colloid on the surface of a solid substrate under the action of continuous laser radiation of different power (up to 10W) when a certain strategy (4) of scanning a laser beam in a colloidal system took place.

This makes it possible to obtain histograms for the size distribution of nanoparticles with the required distributions, including bimodal and trimodal (see Figure 4).

I) II)

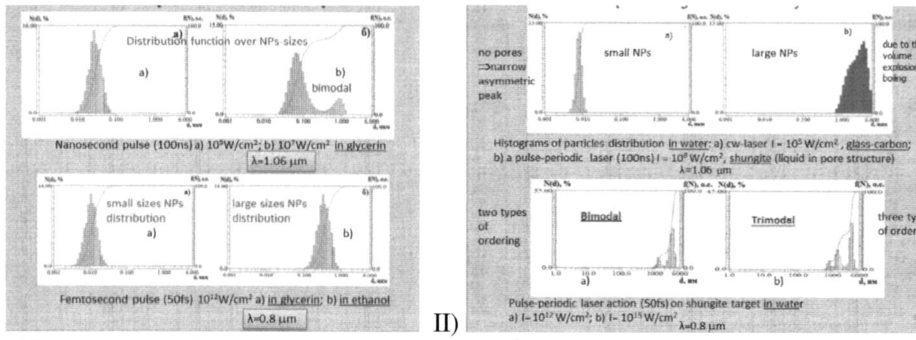

Figure. 4. Histograms of the controlled size distribution of Cu-nanoparticles (I) and carbon-nanoparticles (II) in the colloid for different conditions (they are shown directly in the Figures).

The fundamental features of the nanoclusters under consideration (see Table 1) are determined by the following factors [12].

(1) Physical meaning – is determined by not only the number of atoms N in cluster but number of electronic states, i.e. (N·valence) – valent electrons resulting in an effective nanocluster size.

(2) The ratio of the number of atoms on the surface layer (of the volume ΔV with the surface thickness δ) – N_δ to the number of atoms in the volume V – N_v is comparable $N_\delta/N_V \sim 1$. Thus, new physics arises due to individual interaction of both surface and bulk particles.

(3) The Core – Shell model, and therefore the magic numbers of particles can be applied to the nanocluster object under some conditions.

Table 1. Number N of nanoparticles in clusters (the size d) for different substance

Substance	C	Fe	Ag	W	Au
Density, g/cm^3	3.5	7.9	10.5	19.4	19.3
M $_{\text{atom unity mass}}$	12	55.8	108	184	197
N (d =100 nm)	$9.1*10^7$	$4.25*10^7$	$2.9*10^7$	$3.1*10^7$	$2.9*10^7$
N (d =20 nm)	$7.3*10^5$	$3.4*10^5$	$2.3*10^5$	$2.5*10^5$	$2.3*10^5$
N (d =10 nm)	$9.1*10^4$	$4.25*10^4$	$2.9*10^4$	$3.1*10^4$	$2.9*10^4$
N (d =5 nm)	$1.1*10^4$	$5.3*10^3$	$3.6*10^3$	$3.9*10^3$	$3.6*10^3$
N (d =2 nm)	$7.3*10^2$	$3.4*10^2$	$2.3*10^2$	$2.5*10^2$	$2.3*10^2$
N (d =1 nm)	91	42	29	31	29

If we are talking about the peculiarities of solid state physics, the goal is to change the functional properties of the units. The physical properties of nanocluster systems are very sensitive to the form, size and distance between their composing elements, and the fact is very well known for any material in general, but to change these parameters and to carry out the stable conditions for ordinary solid state object we need both to put the object under extremal high pressure ($\gtrsim 10^6$ atm) and to work in low (liquid He) temperature range ($\lesssim$30K) [13].In contrast, nanocluster structures can be easily modified in necessary direction and also by controlled way in femto- nanophotonics experiments [9]. The variation of the enumerated above topology parameters can result in new type of correlation states for charged particles with coupled states. Moreover, the electronic energetic bands of the materials can dramatically vary in the case, resulting in new physical behavior of the system.

We completed several laser procedures for obtaining nanostructures and thin films with controllable topology. They occur under the development of different nonlinear processes in the system (thermodiffusion, gas-dynamic evaporation in pore-like structures with bubbles, ablation products, ballistic movement of the particles in liquid). All these processes can be analyzed in accordance with different nonlinear hydrodynamic regimes. We discuss both some possible nonlinear mechanisms being responsible for a high electroconductivity and the features of obtaining a hopping conductivity in such inhomogeneous the thin film surface structures (with the thickness of up to 100 nm) when the charged particles are propagating along their boundary surface.

A simple 2-steps mechanism for enhancement of quantum behavior (e.g. in electroconductivity) exists for different conditions.

First, when inelastic length $l_{inelastic} > a_{cluster}$ (where $a_{cluster}$ – the cluster size) we have not incoherent electron-phonon (*e-ph*) scattering, and coherent process occurs with spatial scale ℓ_{coh}.

Second, when de Broglie length $\lambda_{dB} \equiv \ell_{coh} > \Lambda$ (Λ – spatial period of the nanoparticle distribution) the coherent tunneling without loss occurs, and a long-range order with interference of the states takes place in the medium due to lattice/grid structure in the system.

Under such conditions, the electroconductivity enhancement can result in paradoxical phenomenon when electroconductivity in granulated structure may be higher than in monolith sample due to many surface/boundary units – like topological states in topological insulator [1, 4].

Moreover, in terms of basic aspects we can discuss a steady state random ergodic processes in cluster dynamic systems under conditions of dimensional quantization. The problems arise in two aspects.

First, a breaking of ergodic state due to several key-factors being specific: initial state principal item, coupling processes, dimensional quantization.

Second, optimal topology states can be induced with paradoxes in probability theory and statistics when a simple presentation for ergodic dynamic stochastic process does not work: final probability distribution due to averaging over time t does not equivalent to averaging over statistical spatial ensemble.

In the case, continuous spectrum for development of the unit in time and dissipation can result in strange attractor. In fact, it is possible to carry out the key parameter modelling for hopping phenomena by selected cluster energy in a quantum evolution system, and we have non-ergodicity behavior in electroconductivity in granular nanostructure due to the enhancement contribution effect for several states (cf. [14]) being the new phase states.

Let's consider the 1D-structures and functional characteristics of different types for the case. The principal modeling results are presented in Figures. 5,6 (cf. [15]).

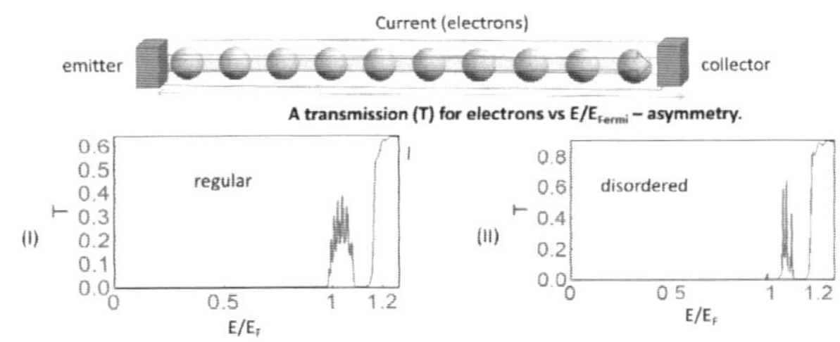

Figure 5. Electroconductivity in a quasi 1D-case. Electric quantum tunneling through 10 nanoclusters (Kronig-Penney 1D-model), i.e. nanocluster array (nanocluster superlattice) in both ordered (I) and disordered cases (II).Transmittance of a quasi 1D-structure consisted of nanoclusters as a function of the electron energy E (by ratio E/E_F, were E_F – Fermy energy) for the regular (left) and disordered (right) structure. The numerical values: N_e =40, r_s =2.1, m_{eff} =1.4, m_0 =0.708MeV, $E_F \approx$ 6eV $\hbar/\tau=10^{-3}E_F$, $\ell \approx$20nm (distance between two clusters). Here r_s – $Wigner\text{-}Seitz\ parameter \left(\sim \frac{1}{N_e^{\frac{1}{3}}} \right)$, $r_s = 2.1$.

Quantum-mechanical tunneling of electrons can occur both (1) into the nanocluster and (2) from the nanocluster into a neighbored nanoclusters. Resonant tunneling refers to tunneling in which the electron transmission coefficient through a structure is sharply peaked about certain energies. For electrons with an energy corresponding approximately to the virtual resonant energy level of the quantum well, the transmission coefficient is close to unity. That is why an electron with this resonant energy can cross the potential barrier without being reflected. This resonant phenomenon is similar to that taking place in the optical Fabry-Perot resonator and/or in the microwave capacitive-coupled transmission-line resonators, i.e. like Klein passing without reflection (see e.g. [16]).

The e-transmission spectrum of superlattice is splitted into a number of spectral bands (like Fano-resonance/interference).For calculation, we used transfer matrix approach with the Morse potential $V(r) = D_e\left(1 - e^{-a(r-r_0)}\right)^2$, where D_e – a depth of the potential well; $a = \omega_0\sqrt{m_e/2D_e}$ – the width of the potential well; ω_0 – a frequency of harmonic oscillator; r_0 – a local minimum position (for equilibrium state) [1].

As soon as the voltage at the microcontacts of the quantum well structure (see Figure 5) is out to corresponding energy of one of its resonant states, the tunneling current increases (region 1).

With a positive bias applied to the right contact relative to the left, the Fermi energy on the left is pulled through the resonant level E_F. As the Fermi energy passes through the resonant energy, a large current flows occur due to the increased transmission from left to right. It leads to significant increase of current (region 2). With higher bias, the current ceases to flow, when E_F falls below the conduction-band edge. The result is a marked decrease of the current with increasing voltage, giving rise to a region of negative differential resistance or significant current suppression (region 3 – in accordance to the plateau in Figure 5). At larger bias the current increases again as particles acquire enough kinetic energy (region 4). Electrical current J is defined as: $J = eS\int_0^{p_f}\frac{\hbar k_z}{m^*}T(k_z)\frac{2\pi\,(p_F^2-\hbar^2k_Z^2)}{(2\pi\hbar)^3}\hbar dk_z$, where S – the cross-section area of the structure, p_F – Fermi momentum, $T(k_z)$ – transfer matrix for transmission coefficient.

A special case of 1D-structure is a creation of Long Linear Carbon Chain (LLCC) being a new carbyne phase for carbon [17].

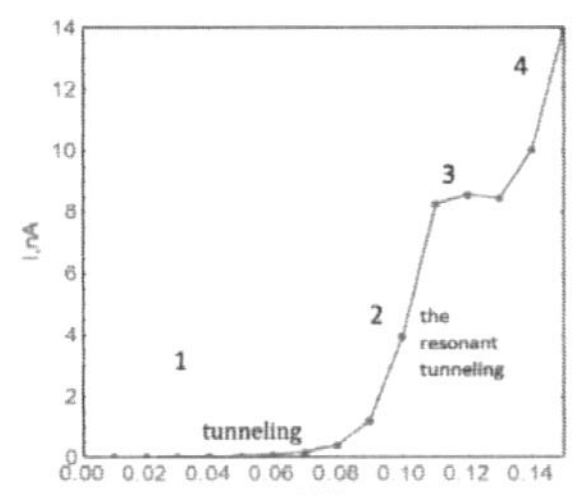

Figure 6. Voltage-current characteristic (VAC) of a quasi 1D-structure consisting of nanoclusters (the explanations are given in the text).

As an example we introduce one of our experimental results for Volt-Ampere Characteristics (VAC) for complex Au/Ag with carbon, resulting in a LLCC (see [18]), that presented in Figure 7.

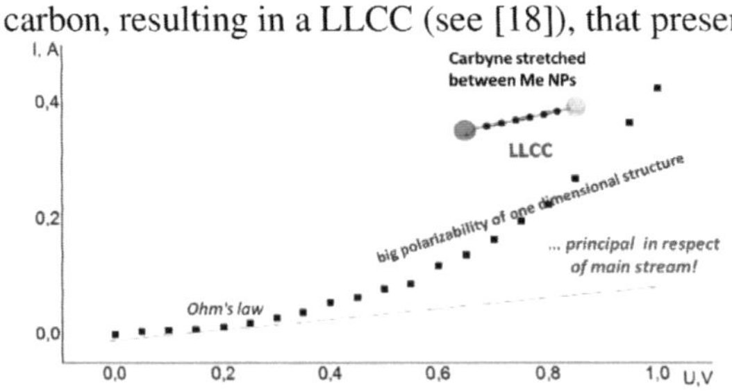

Figure. 7. VAC of the Au/Ag-Carbone linear structure (i.e. Au/Ag-carbyne structure) thin films with thickness 30nm on glass substrate. The enhancement of electroconductivity in comparison with the linear Ohm's dependence can explain by a big polarizability of linear dimensional 1D-structure – LLCC.

This result is principal because demonstrates the possible trend with comparison with the behavior like tendency to superconductivity in the 1D-electron stripe-stretch structure of cuprates O+Cu (they discuss high temperature superconductivity up to 600°C (!)) – cf. [19, 20].

All these data result in new states in the topology photonics and optoelectronics.

In tact, for nanostructures (granulated size 10-100 nm) the thermodynamic, kinetic/transport and structure properties depend on the composite size particles due to their chemical nature, morphology of the particles, interphase surface and some interaction dimension effects like in heterogeneous systems. And we can speak about a real phase transition (vs thermodynamic parameters) but in this case not vs temperature T (like in condensed matter) and also both not vs external fields (e.g. vs magnetic field H in Liquid Crystal) and vs laser pulse duration (the temperature phase transition trajectory in fs-time domain) but now with a principal new behavior, i.e vs particle size – the nanostructure topology.

Physical reason for last case is that the inhomogeneous structure results in self-overcome of the energetic barriers for equilibrium states; in non-equilibrium systems the velocity dependence of the process in inhomogeneous structure is determined by kinetic parameters of the unit.

Finally, in general for clearance, the laser ablation (used by us) diagram in both time and space domains is shown in Figure 8 [21, 22].

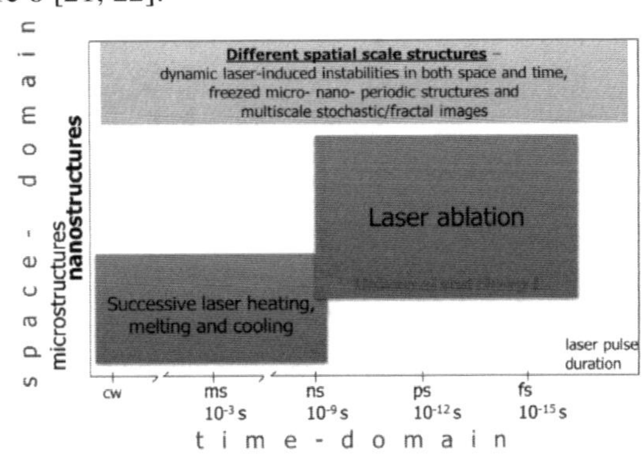

Figure. 8. Laser landscape – the 4D-technologies map [21,22].

3. Dendrite/fractal structures – prototypes of future topological elements

The model DLA – Diffuse Limited Aggregation of deposition and the formation of complex dendrites/fractal structures on surface, is based on the procedure with the algorithm consists of the several steps [23]:

1) on a square two-dimensional lattice, a germ structure is specified;
2) far from the cluster (from the nucleus), a new particle is generated in the nucleation region;
3) a new particle wanders randomly, taking into account the viscosity of the medium;
4) if a particle comes to a busy cell, then it sticks with a given probability;
5) if the particle moves far enough away from the cluster, it is destroyed;
6) repeat, starting from step 3, until the particle sticks with a given probability, after which a new particle is launched.

In the computer simulation we used both the initial concentration of the aggregation centers varied (center and lower boundary — (a), (b); and the probability of sticking of particles within the two-dimensional von Neumann neighborhood of order 1 (c) – Figure 9.

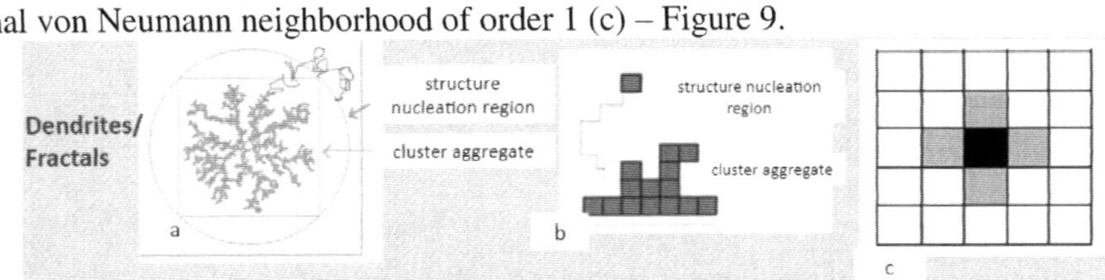

Figure. 9. Island model – DLA-scheme (Diffuse Limited Aggregation): (a) – formation from the center of the computational domain; (b) – formation from the lower boundary of the computational domain; (c) – von Neumann neighborhood.

The results of both calculation and experiments are presented in Figures 10, 11.

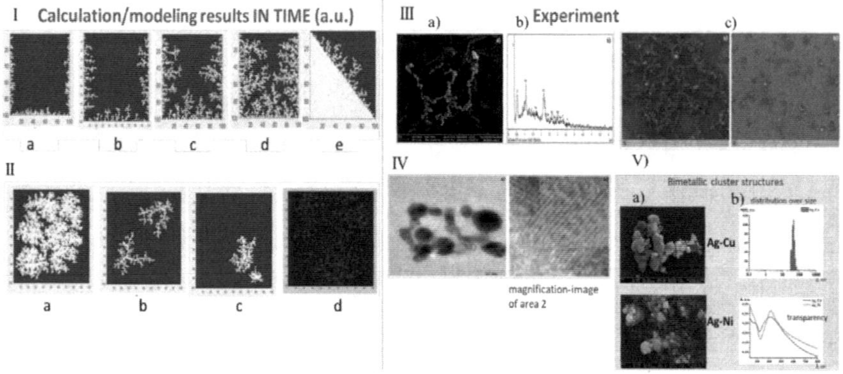

Figure 10. Development of Dendrite/Fractal structures.

I. Aggregation centers (top view) – the line on both the side and bottom boundaries for different islands of the computational domain: p = 0.1, sc = 0.1; (b) p = 0.1, sc = 0.5; (c) p = 0.5, sc = 1; (d) p = 0.8, sc = 1; (e) p = 0.5, sc = 1, where the controlling parameters are: sc – particle sticking probability (vs diffusion) ; p – viscosity.
II. The Model films (below view): 5 islands: p = 1, sc = 0.1 (a); 2 islands: p = 0.1, sc = 1 (b); 2 islands: p = 0.1, sc = 0.1 (c); 5 islands: p = 0.5, sc = 1 (d).
III. REM-images of deposited Au-Ag cluster on the carbon surface (a) and X-ray spectra of the elements (b). REM-images of Au-Ag cluster deposited on the quartz surface with various concentration of nanoparticles in water (c): 1:10 (left); 1:20 (right).
IV. TEM-image of deposited Au-Ag cluster (area 1), lattice constants d_1=2,05±0,07A; d_2=2,07±0,07A (area 2). The interplanar values d(002) for both Au, and Ag.
V. Bimetallic cluster structure: (a) – structure, (b) – the particle size histogram (up) and transperancy spectrum (down).

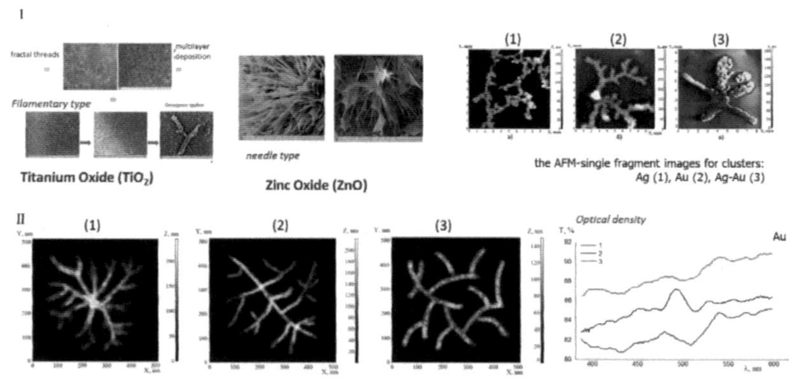

Figure 11. (I) Obtained different Fractal structures in Experiment (up platform). (II) Morphological variety of the deposited cluster and consequent optical density (down platform). Explanations are directly given in the Figures.

In progress, such complex structures may establish a new mechanism of the charge transport in thin films based on coherent migration of the deformation of the nanoclusters potential, which carries with it an electron. We define a defect in the form of a potential deformation, similarly to a defection (cf. e.g. [24], and see also [1,4,10]), as a quasiparticle being a distortion.

Conclusion

Let's enumerate the summary collections over our different experiments on the topology peculiarities of the granulated metallic film deposited on dielectric substrates – in frame of the topology photonics (cf. [7,25]).

First, in electroconductivity:
- the tunneling delocalized effect depends on the size of NPs, distance between them and shape of tunnel barriers and also vs variation of the thickness of the films, transition from amorphous to crystalline structure;
- the thermally activated hopping regime of the electron transport between localized centers;
- the topology enhancement electroconductivity tendency for cluster like for topological insulator with coupling electron states.

Second, in optical characteristics:
- the broadening of energetic electronic levels occurs;
- optical metasurfaces characteristics arise;
- strong optical response occurs;
- density of e-state (on Fermi-surface) can be varied by controlling way.

Third, the symmetry breaking process (Majorana states for quasi-particles, the quantum Hall partial size effect, Landau-levels, Tamm-states …) may be developed.

The principal problem on the subject is how to determine specific numerical values of the topology control parameters of a nanocluster system relying on these general relations for the achievement of the desired final result in the experiment. This requires a detailed research (in both theory and experiment) of the influence of nanocluster topological parameters on the functional properties of the substances used.

Nevertheless, the discussed phenomena give us an opportunity to establish the basis of new physical principles to create the functional elements for topological photonics in hybrid set-up (optics + electrophysics) being controlled by quantum coupled states and nonlinear dynamic processes. In progress, we can discuss the tendency and trends to superconductivity by adjustable both the element composition and the topology structure (cf. [26,27]).

Finally, we enumerate the list of perspectives-accents hybrid circuits (electrophysics + optics) [21].

A. Nanoelectronics and Materials: 2D materials, such as transition metal dichalcogenides (for example, MoS_2), black phosphorus (in some sense similar to graphene), topological insulators, and others.

Key parameters.

(1)Transport properties: topological insulators, low scattering and high mobility of current carriers.

(2)Power dissipation and scaling.

Directions.

(a) Neuromorphic computing, i.e. parallel analog computing, such as neural networks.

(b) Spintronics.

B. Nanophotonics: terahertz modes, nanoplasmonics and metamaterials.

C. Nanoenergy: nanostructured systems for improved light collection.

D. Nanomedicine: nanomaterials for the development of cell/molecular biology and tissue engineering.

In conclusion, let's enumerate the few points of modern advantages and activities in this several selected directions:

(1) A squeezed quantum microcomb on a chip, [28].
The quantum computing market is projected to reach $65 billion by 2030, a hot topic for investors and scientists alike because of its potential to solve incomprehensibly complex problems.
(2) Single-electron spin resonance in a nanoelectronic device using a global field by [29].
(3) Preparation and Readout of Multielectron High-Spin States in a Gate-Defined GaAs/AlGaAs Quantum Dot [30].
(4) Quantum proton entanglement on a nanocrystalline silicon surface [31].
(5) Layer Hall effect in a 2D topological axion antiferromagnet. [32].
(6) A post-quantum chip with hardware Trojans [33].

Acknowledgments
The reported study was funded by RFBR, project number № 20-02-00515, № 20-32-90052 and the Ministry of Science and Higher Education of Russian Federation within the State assignment № 0635-2020-0013.

The research was carried out using the equipment of the interregional multidisciplinary and interdisciplinary center for collective use of promising and competitive technologies in the areas of development and application in industry/mechanical engineering of domestic achievements in the field of nanotechnology (agreement No. 075-15-2021-692 of August 5, 2021).

References

[1] Gantmacher V F 2013 *Electrons in disordered media* (Moscow: Fizmatlit) p 232
[2] TSMC will increase investment in chip production in the United States from $12 billion. up to $35 billion. [Internet resource]//[official website] https://hightech.plus/2021/03/14/tsmc-uvelichit-investicii-v-proizvodstvo-chipov-v-ssha-s-12-mlrd-do-35-mlrd (accessed 15.03.2021)
[3] Apple and TSMC are working together on a 2-nm processor [Internet resource]//[official website] https://hightech.plus/2021/03/14/apple-i-tsmc-sovmestno-rabotayut-nad-2-nm-processorom (accessed 15.03.2021)
[4] Dragunov V P, Neizvestny I G and Gridchin V 2006 *A Fundamentals of nanoelectronics: A textbook* (Moscow: Logos) p 496
[5] Scientists have assessed the future of photonic computers from the point of view of the latest technological achievements – and it looks very promising [Internet resource]//[official website] https://news.myseldon.com/ru/news/index/244848012 (accessed 13.08.2021); N. D. Rizzo. Nanoelectronics to improve energy efficiency//Everspin Technologies. 2009. 11. Access mode: https://www.nist.gov/system/files/documents/2017/05/09/228_nanomaterials_nanotechnology_nanoelectronics.pdf.
[6] Galekovic G N, Severtsev V N and Shurchkov I O 2012. *Engineering Bulletin of the Don* **2(20)** 315-319
[7] Arakelian S M, Chestnov I Yu, Istratov A V, Khudaiberganov T A and Butkovskiy O Ya 2020 *Nonlinear dynamic modeling for high temperature superconductivity in nanocluster topological structures on solid surface* (New Trends in Nonlinear Dynamics. Proceedings of the First International Nonlinear Dynamics. Proceedings of the First International Nonlinear Dynamics Conference (NODYCON 2019) Springer, Cham 3) 121-130
[8] NS RAS "Quantum technologies" [Internet resource]// [official website] http://www.ras.ru/news/shownews.aspx?id=d96f9b47-49cb-41f8-8627-3f6e3c6e0d06#content (accessed 16.03.2021)
[9] Arakelian S M, Kucherik A O, Prokoshev V G, Rau V G and Sergeev A G 2015. *Introduction to femtonanophotonics. Fundamental principles and laser methods of controlled production and diagnostics of nanostructured materials: a textbook* (Moscow: Logos) p 744
[10] Ignatov A N 2012 *Nanoelectronics. State and prospects of development: Training manual.*

Manual. (Moscow: Flint) p 360
[11] Arakelian S, Kutrovskaya S, Kucherik A, Emel'yanov V and Zimin S 2016 *Optical and Quantum Electronics* **48** 342 1–16
[12] Magunov A N 2012 *Spectral pyrometry* (Moscow: Fizmatlit) p 248
[13] Kresin V Z and Ovchinnikov Yu N 2008 *Physics-Uspekhi* **51(5)** 427–435
[14] Faoro L, Feigel'man M V and Ioffe L B 2018 *Annals of Physics* **409(8)** 167916
[15] Arakelian S, Bukharov D. Khudaiberganov T and Osipov A 2021 *Journal of Physics: Conf. Series* **1822** 012011
[16] Demikhovsky V and Vugalter G 2000 *Physics Of Quantum Low-Dimensional Structures* (Moskow: Logos) p 250
[17] Kaiser K and Scriven L 2019 *Science* **10** 1126
[18] Kutrovskaya S V, Arakelian S M, Kucherik A O, Osipov A V and Garnov S V 2019 *Laser Physics* **29(8)** 627–33
[19] Huang E W, Mendl C B, Liu S, Johnston S, Jiang H C, Moritz B and Devereaux T P 2017 *Science* **358(6367)** 1161–64
[20] Wang F, Kivelson S and Lee D H 2015 *Nature Physics* **11** 959–63 (arXiv preprint 1501.00844)
[21] Gudnik S 2017 *Nano and Giga Challenge in Electronics, Photonics and Renewable Energy* (Arisona State University) p 326
[22] Yang G 2012 *Laser Ablation in Liquids: Principles and Applications in the Preparation of Nanomaterials* (Pan Stanford Publ. press LTD) p 1192
[23] R Kobayashi 1993 *Physica D: Nonlinear Phenomena* **63(3-4)** 410–23
[24] Sedov E, Sedova I, Arakelian S, Eramo G and Kavokin A 2020 *Scientific Reports* **10 (1)** 8131
[25] Kavokin A, Kutrovskaya S, Kucherik A, Osipov A, Vartanyan T and Arakelian S 2017 *Superlattices and Microstructures* **111** 335–39
[26] Kutrovskaya S, Osipov A, Baryshev S, Zasedatelev A, Samyshkin V, Demirchyan S, Pulci O, Grassano D, Gontrani L, Hartmann R R, Portnoi M E, Kucherik A, Lagoudakis P G and Kavokin A 2020 *Nano Letter* **20(9)** 6502–09
[27] Yugai K N 2013 *Bulletin of OmSU* **2(68)** 104-7
[28] Yang Z, Jahanbozorgi M, Jeong D, Sun S, Pfister O, Lee H and YiA X 2021 *Nature Communications* **12(1)** 4781
[29] Vahapoglu E, Slack-Smith J P, Leon R C C, Lim W, Hudson F E, Day T, Tanttu T, Yang C, Laucht A, Dzurak A S and Pla J J 2020 *Science Advances* **7(33)** 9158 (arXiv preprint arXiv:2012.10225)
[30] Kiyama H, Yoshimi K, Kato T, Nakajima T, Oiwa A and Tarucha S 2021 *Phys. Rev. Lett.* **127** 086802
[31] Matsumoto T, Sugimoto H, Ohhara T, Tokumitsu A, Tomita M and Ikeda S 2021 *Phys. Rev. B* **103** 245401
[32] Gao A, Liu YF and Hu C 2021 *Nature* **595** 521–5
[33] A post-quantum chip with hardware Trojans [Internet resource]// [official website] https://www.tum.de/en/about-tum/news/press-releases/details/36835/ (accessed 04.08.2021)

Terahertz technology in diagnosis of glioma molecular markers

S Kuznetsov[1,2], M Konnikova[3,4], T Heinz[3,4], E Dizer[5], N Nikolaev[6,7], D Utkin[1,2] and O Cherkasova[4,7]

[1]Physics Department, Analytical and Technological Research Center, Novosibirsk State University, 2 Pirogova St., Novosibirsk 630090, Russia
[2]Rzhanov Institute of Semiconductor Physics SB RAS, 13 Academician Lavrentiev Ave., Novosibirsk 630090, Russia
[3]Department of Physics, Lomonosov Moscow State University, 1 Leninskiye Gory, Moscow 119991, Russia
[4]Institute for Problems of Laser and Information Technologies RAS - Branch of the FSRC «Crystallography and Photonics» RAS, 1 Svyatoozerskaya St., Shatura 140700 Moscow Region, Russia
[5]National Research Nuclear University "MEPhI", 31 Kashirskoe shosse, Moscow 115409, Russia
[6]Institute of Automation and Electrometry SB RAS, 1 Academician Koptyug Ave., Novosibirsk 630090, Russia
[7]Institute of Laser Physics SB RAS, 15 B Academician Lavrentiev Ave., Novosibirsk 630090, Russia

E-mail: o.p.cherkasova@gmail.com

Abstract. Gliomas are invasive brain tumors with high rates of recurrence and mortality. It has been shown that specific markers for glioma's differential diagnostics are enantiomers of 2-hydroxyglutarate (L-2HG and D-2HG) in brain tissues and blood. These isomers have unique absorption peaks originating from vibrational and rotational transitions in their molecules. In particular, the peaks centered at 1.337 THz and 1.695 THz correspond to L-2HG and D-2HG isomers, respectively. The goal of this work is to develop highly efficient frequency-selective sensors for L-2HG and D-2HG isomers using the effect of nanoantenna-assisted plasmonic enhancement of THz absorption. Such an approach provides a noticeable increase in detection sensitivity versus direct non-resonant methods. In this paper, we present the numerical results of the design optimization for L-2HG and D-2HG sensors based on Si/SiO_2-wafer-backed arrays of golden nanoantennas of linear geometry. The optimal structural parameters of the arrays found through integral averaging of the square of the surface electric field over an array unit cell are recommended for further nanolithographic fabrication of this kind of THz sensor.

1. Introduction

Glioma is one of the most common tumors in the central nervous system [1]. Its identification at the earliest stages is a medical task of primary importance [2]. Among molecular biomarkers of gliomas, the mutated enzyme of isocitrate dehydrogenase 1/2 (IDH1/IDH2) was recently identified [3]. IDH1/2 mutations cause accumulation of D and L enantiomers of 2-hydroxyglutarate (2HG) at the 10-100-fold higher level than that in IDH wild-type tumors or healthy tissues [4]. In gliomas, 2HG is a mixture of L-2HG and D-2HG whose relative content distinguishes glioma against acute myelogenous leukemia: the content of L-2HG is found to be larger than that of D-2HG for the glioma case, respectively [5].

D- and L-isomers have unique THz absorption peaks originating from vibrational and rotational transitions in their molecules [6]. In particular, the peaks centered at 1.337 THz and 1.695 THz correspond to L-2HG and D-2HG isomers, respectively. The final goal of this work is to develop theoretically and experimentally the highly efficient frequency-selective sensors for these substances using the effect of nanoantenna-assisted plasmonic enhancement of THz absorption. This effect, serving as a basis for nanoantenna-enhanced terahertz spectroscopy (NETS) [7] and considered as a kind of surface-enhanced infrared absorption (SEIRA) [8], is based on many-fold resonant amplification of the electromagnetic field in local proximity of nanosized metallic elements at the frequency of their localized surface plasmon resonance, which is chosen to coincide with the energy transition frequency of molecules under detection. The high efficiency of this technology versus direct non-resonant methods in detecting analytes of extremely low concentrations was demonstrated in different issues, e.g. recognition of sugars [9] and steroid hormones [10]. We used such a technique for the detection of the cortisol hormone in human blood plasmas [11, 12].

It is worth highlighting that NETS exploits Fano-like interference between the fundamental mode of nanoantennas and the characteristic resonance of analyte molecules. This effect is proportional to the square of the electric field in the proximity of the nanoantennas and does not shift their plasmon resonance in the low-concentration regime. It enables detecting single molecular layers and distinguishes NETS from the "more classical" and less sensitive approach used in THz metamaterial sensors with an analyte-induced spectral shift of the plasmon resonance caused by changing the dielectric environment of the metamaterial particles [13-17].

In this paper, we presented the selected results of the design optimization for THz sensors based on Au nanoantenna arrays (NAs) supported by Si/SiO_2-substrates. The considered NA sensors have linear geometry and are designed for the resonant frequencies of 1.337 and 1.695 THz corresponding to L-2HG and D-2HG isomers, respectively. Complete work with a description of the experimental implementation for these sensors using E-beam lithography [11, 18, 19] and their further characterization with a manufactured-in-house THz-TDS system [17] will be published elsewhere.

2. Problem Definition

The geometry of NAs under study is illustrated in Figure 1. The nanoantennas represent highly elongated in X-direction linear bars of gold (Au) which are arranged regularly on the surface of a high resistivity Si wafer with a uniform SiO_2 overlayer. Specified by our fabrication technology, the Au thickness t, and butt-to-butt spacing A between nanoantennas were fixed as 50 nm and 100 nm, respectively, while the available SiO_2 thickness h_{SiO2} was considered in the range of 0-100 nm. The linear geometry of nanoantennas and their large inter-element distance in Y-direction were intentionally chosen to maximize the quality factor and, therefore, the local electric field of nanoantennas at the frequency of their fundamental dipole resonance.

The NAs of the proposed configuration were numerically investigated in ANSYS Electromagnetics Suite R19 software using the regime of Floquet ports and periodic boundary conditions applied to the NA unit cell. The silicon wafer was considered as a semi-infinite in –Z-direction that was realized by placing Si on the bottom Floquet port. Gold was modeled as a Drude medium with the linear plasma and damping frequencies of 72500 cm^{-1} and 216 cm^{-1}, respectively [18, 19]. For realistic imitation of the manufactured structures, the nanoantennas were modeled as rectangular bars with chamfered edges (from the vacuum side) having the fillet radius of 15 nm.

The goal of the full-wave simulations was to determine the optimal geometric parameters of nanoantennas, such as their length L and width B, transverse lateral pitch G_y, and SiO_2 thickness h_{SiO2} under conditions of keeping the NA resonance at ν_{res} = 1.337 THz (L-2HG) or 1.695 THz (D-2HG) and maximizing the square of the normalized surface electric field $\langle E^2 \rangle$ averaged over the NA unit cell within the nanoantenna height (see also [19]):

$$\langle E^2 \rangle\,|_{\mathrm{v}=\mathrm{v}_{res}} = \frac{\int_0^t \int_{-\frac{1}{2}G_y}^{\frac{1}{2}G_y} \int_{-\frac{1}{2}G_x}^{\frac{1}{2}G_x} |\boldsymbol{E}(x,y,z)|^2 \, dx\, dy\, dz}{\int_0^t \int_{-\frac{1}{2}G_y}^{\frac{1}{2}G_y} \int_{-\frac{1}{2}G_x}^{\frac{1}{2}G_x} dx\, dy\, dz} \tag{1}$$

In (1), $\boldsymbol{E}$ is understood as the field normalized to that on the surface of bare Si without NA. The integral averaging with normalization to the unit cell volume described by the denominator is necessary to adequately compare the sensing performance of NAs with different geometries.

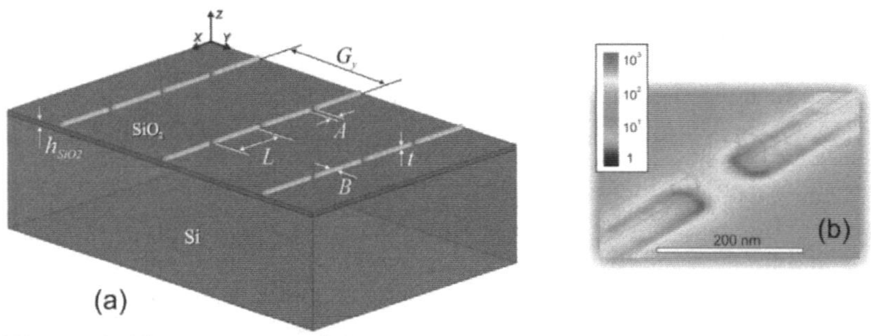

Figure 1. The geometry of the Au nanoantenna array (a) and illustration of the surface electric field $\boldsymbol{E}$ enhancement at the plasmonic resonance (b). $\boldsymbol{E}$ is normalized to the field on the bare Si surface without NA. The operating wave polarization is parallel to nanoantennas ($\boldsymbol{E}$ || X).

3. Selected simulation results

Figure 2 (a) shows typical spectra of the wave transmission from vacuum to Si in terms of $|S_{21}|^2$ parameter under normal X-polarized excitation. The curves correspond to the example of SiO_2-free NAs with B=100 nm and optimal L and G_y values yielding the plasmonic resonances at 1.337 THz and 1.695 THz. For NA sensors of this kind the resonances have a dipole nature and are manifested as the dips in transmission or peaks in reflection.

Figure 2 (b)-(d) exemplify the results of parametric simulations for the L-2HG sensor, where the optimal nanoantenna length L and the integral intensity amplification factor $\langle E^2 \rangle$ derived from the formula (1) using the ANSYS HFSS field calculator are depicted as the functions of G_y, B, and h_{SiO2}. The presented data allow one to draw the following conclusions:

- For any NA the Q-factor of the plasmonic resonance and, therefore, the $\langle E^2 \rangle$-factor grow upon increasing the inter-antenna spacing G_y until the diffraction lobes appear in the NA substrate (Figure 2(b)). For the rectangular arrangement of NA unit cells and normal illumination, the diffraction lobes onset point corresponds to the wavelength condition

$$G_y = \lambda_{res}/n_{Si}, \tag{2}$$

  where n_{Si} is the refractive index of silicon. In practical scenarios, it is recommended to use G_y slightly smaller than the value calculated from (2) to avoid crossing the diffraction point above which the $\langle E^2 \rangle$-factor sharply decreases. For the L-2HG and D-2HG sensors the values G_y = 63500 nm and 50000 nm, respectively, can be chosen as optimal.
- There is no sense to use very narrow nanoantennas to enhance the peak near-field at the nanoantenna edges as the relative area of the enhancement zone may be integrally too small. By calculating $\langle E^2 \rangle$, the optimal nanoantenna width B exhibiting maximal $\langle E^2 \rangle$ can be determined (Figure 2 (c)). For the L-2HG and D-2HG sensors, this corresponds to the values B = 630 nm and 460 nm, respectively.

- The Q-factor of the plasmonic resonance and the $\langle E^2 \rangle$-factor increase when decreasing the refractive indices of materials in the NA dielectric environment. This allows recommending presence of a SiO_2 layer beneath nanoantennas with a thickness as big as possible (Figure 2 (d)). In our case, the maximally available h_{SiO2} value is 100 nm at which the $\langle E^2 \rangle$-factor can be enlarged by 2-2.5 times against NAs on bare Si. Our simulations also showed that, due to strong localization of the plasmonic field, the regime h_{SiO2} = 100 nm is close to saturation: the $\langle E^2 \rangle$-factor is only 20-25% smaller as compared to the marginal case of the semi-infinite SiO_2 substrate.

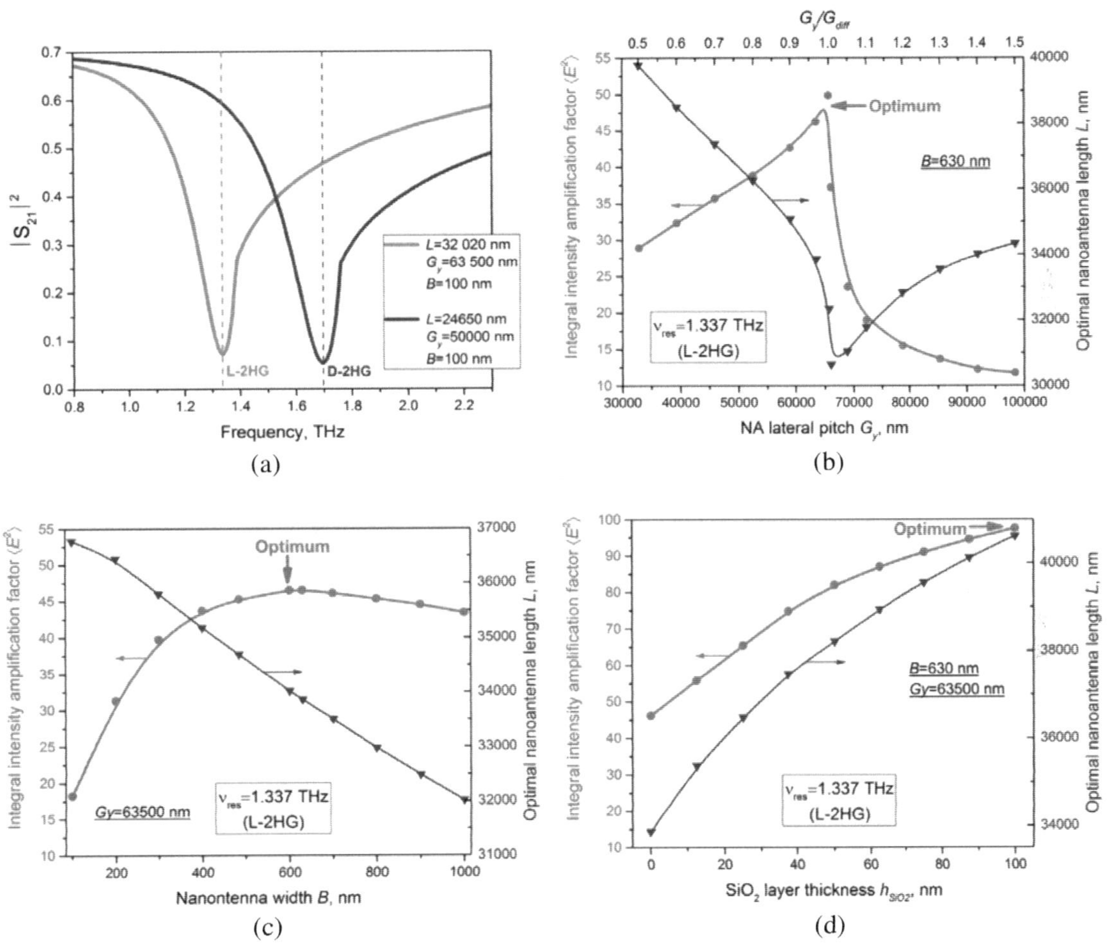

Figure 2. Examples of the simulated transmission spectra for Si-backed SiO_2-free NAs (a) and simulated parametric $\langle E^2 \rangle$- and L-curves for the case of the L-2HG sensor showing the route for choosing its optimal geometric parameters (b-d).

4. Conclusion

The $\langle E^2 \rangle$-factor quantity proposed in this work is evaluated to be the most adequate tool to assess the performance of any NETS sensors. With this tool, we have determined the optimal length L, width B, and SiO_2 sublayer thickness h_{SiO2} for arrays of linear nanoantennas patterned on Si substrates adapted for characteristic frequencies of L-2HG and D-2HG isomers. The full set of parameters for optimized NAs is summarized in Table 1.

Table 1. Optimized structural parameters of Au NA sensors.

Isomer	Resonant frequency THz	Resonant frequency cm^{-1}	Longitudinal pitch $G_x=L+A$, nm	Transverse pitch G_y, nm	Nanoantenna length L, nm	Nanoantenna width B, nm
L-2HG	1.337	44.60	40 720	63 500	40 620	630
D-2HG	1.695	56.54	33 269	50 000	33 169	460

Currently, the developed nanoantenna sensors have passed through fabrication and are at the stage of in-situ testing to quantify their real sensitivity to L-2HG and D-2HG isomers in biological samples.

Acknowledgments
The work was supported by the Russian Foundation for Basic Research RFBR and National Natural Science Foundation of China, research project № 19-52-55004; the Ministry of Science and Higher Education of the Russian Federation within the State assignment FSRC "Crystallography and Photonics" RAS and project № 121032400052-6; the Interdisciplinary Scientific and Educational School of Moscow University “Photonic and Quantum Technologies. Digital Medicine”.

References
[1] Ostrom Q T, Cioffi G, Gittleman H, Patil N, Waite K, Kruchko C and Barnholtz-Sloan J S 2019 *Neuro-Oncology* **21**(Suppl 5) v1–v100.
[2] Cherkasova O, Peng Y, Konnikova M et al 2021, *Photonics,* **8** (1) 22.
[3] Waitkus M S, Diplas B H and Yan H 2016 *Neuro-Oncology* **18** (1) 16–26.
[4] Dang L, White D W, Gross S et al 2009 *Nature* **462** 739–44.
[5] Strain S K, Groves M D, Olino K L and Emmett M R 2020. *Clinical Mass Spectrometry* **15**, 16–24.
[6] Chen W, Peng Y, Jiang X et al 2017 *Scientific Reports* **7**, 12166.
[7] Piccoli R, Rovere A, Toma A, Morandotti R and Razzari L, “Terahertz Nanoantennas for Enhanced Spectroscopy”, in *Terahertz Spectroscopy - A Cutting Edge Technology*, ed. J. Uddin, IntechOpen (March 13, 2017), DOI: 10.5772/66349. Available from: https://www.intechopen.com/chapters/53269
[8] M. L. de la Chapelle, A. Pucci, ed. *Nanoantenna: Plasmon-Enhanced Spectroscopies for Biotechnological Applications*, CRC Press, Taylor & Francis Group, 2013.
[9] Lee K-D, Kang L-H, Lee J-S et al 2015 *Scientific Reports* **5** 15459.
[10] Lee S-H, Lee D, Choi M H, Son J-H and Seo M 2019 *Anal. Chem.* **91** 6844–6849
[11] Milekhin A G, Cherkasova O P, Kuznetsov S A et al 2017 *Beilstein Journal of Nanotechnology* **8** (1), 975– 81.
[12] Milekhin I A, Kuznetsov S A, Milekhin A G, Duda T A, Cherkasova O P, Rodyakina E E, Latyshev A V 2018 *Journal of Physics: Conf. Series* **1092** 012087.
[13] Xu W, Xie L, and Ying Y 2017 *Nanoscale* **9** 13864-13878.
[14] Lin S, Xu X, Hu F et al 2021 IEEE Journal of Selected Topics in Quantum Electronics **27**(4), 1.
[15] Gu H., Shi C., Wu X. et al 2020 *Analyst* **145**(20) 6705.
[16] Peng Y, Shi C, Zh Y et al 2020 *PhotoniX* **1,** 12.
[17] Nikolaev N A, Kuznetsov S A, Beruete M 2018 *Proc. 48th European Microwave Conf.* pp. 420-423, Madrid, Spain, September 23-27, 2018.
[18] Milekhin I A, Kuznetsov S A, Rodyakina E E et al 2016 *Beilstein Journal of Nanotechnology* **7** 1519–26.
[19] Milekhin A G, Kuznetsov S A, Sveshnikova L L et al 2017 *Journal of Physical Chemistry C* **121** (10) 5779-86.

Features of the formation of bound states of optical phonons in nanostructures – the role of anharmonism

M Arakelyan

Yerevan State University, Yerevan, Armenia

E-mail: marakelyan@ysu.am

Abstract. In this paper, we discuss the features of the formation of bound states of phonons and the role of anharmonism for such a formation in a dimensionally quantized system in comparison with bulk sample.

1. Introduction

The properties of phonons and phonon interactions in nanostructures must be taken into account in the manufacture of quantum devices. The study of phonon interactions in models describing the properties of nanostructures will make it possible to regulate the scattering frequency, the dissipation rate, resonance conditions, the free path of phonons, as well as to control the evolution of the qubit in quantum computers and heterostructures. [1]. An increase in the energy of the electron-phonon interaction leads to a decrease in the transition time between quantum levels in dimensionally quantized systems [2], which increases the speed of the operating units in devices. The question of the formation of bound states of excitation has attracted considerable attention in recent times.

In a thin film the phonon movement in a direction perpendicular to the plane of the film («y» direction) is limited. The effect of such dimensional limitation of phonons leads to a limitation of the phase space of the wave vector, as a result of which scattering of carriers by phonons in bulk samples prevails over scattering of dimensionally limited phonons. This effect is especially pronounced in nanostructures, and is under consideration below. Such questions are under our study in the paper.

2. Bound states of optical phonons

Under the following boundary conditions and the condition of unitarity of the matrix elements ζ of the transformation matrix from variables to the generalized coordinates characteristic of the film we can write (see, for example, [3])

$$\zeta_0 = \zeta_{(N_{3+1})a} = 0, \ \frac{2}{(N_3+1)} \sum \sin(qn) \sin(q'n) = \delta_{q,q'} \tag{1}$$

and we obtain the dispersion Ω equation for optical phonons in the film:

$$\Omega = 2\sqrt{\frac{\gamma}{\mu}[1 - \frac{2\mu}{m_1+m_2} \sin^2 \frac{q^3 a}{2} (1 + \cos^2 \frac{q_3 a}{2})]}; \ \mu = \frac{m_1 m_2}{m_1+m} \tag{2}$$

where γ is the coefficient characterizing the forces of phonon interaction ($\gamma << 1$); the displacements 0 and $a/2$ of the atoms from the equilibrium positions and in the unit cell with the number n are indicated respectively $\zeta_{n_1,1}$, $\zeta_{n_2,2}$, q_3 – the wave vector of the phonon in the direction of y, m_1, m_2 – the masses of the atoms of the unit cell, N_3 – the number of molecular layers in the y direction.

Next, the e2quation for the bound states of phonons in the film is obtained:

$$V(q) = \frac{2}{N}\sum_{m,n} \exp[iq_{\perp}(m_{\perp} - n_{\perp}) \sin(q_3 ma)/ \sin q_3 na) V_{mn},$$
$$n_{\perp} = a_1 n_1 + a_2 n_2 \quad (3)$$

Where $V_{(q)}$ – matrix element of the energy of interaction of particles «m» and «n» in the q representation.

If we take into account the matrix elements of excitation transfer only between neighboring particles, then

$$V_{m,m+1} = V_{m-1,m} = V,\ V_{(q)} = 4V \cos(q_3 a) \cos(q_{\perp} a). \quad (4)$$

After the transition from summation over to integration, the binding energy ε of phonons was found:

$$\varepsilon = \pm 4\sqrt{\frac{A^2}{(N_3+1)^2} + V^2 \cos^2 q_3 a} \quad (5)$$

where A – anharmonicity coefficient.

At $\varepsilon < 0$, bound states of phonons appear in the film for an arbitrarily small one A. Since the anharmonicity constant in the film is greater than in the bulk sample, the influence of the anharmonicity is greater and the levels corresponding to the bound states are located deeper. The matrix element oscillates depending on the film thickness L, and upon transition to a massive sample takes a maximum value A. As the film thickness $L=(N_3+1)a$ decreases, the bound state levels shift down.

3. Bound states of an electron and an optical phonon considering the electron-phonon interaction

We also studied the bound state of an electron and an optical phonon in a thin film, taking into account the electron-phonon interaction, and estimated the binding energy for GaAs.

3.1 Basic assumptions

The main feature of optical phonons is a relatively large value of their frequency with a small value of dispersion. When this condition takes place, the transfer of an electron from node to node occurs with the preservation of the number of phonons, since the matrix elements of all processes associated with a change in the number of phonons enter with large energy denominators. Therefore, the problem changes its character from field-related: «electron + phonon field» to purely dynamic [4].

So, we will subject the Hamiltonian of the «electron + phonon field» system to a canonical transformation and, with an error $\varepsilon_{\varphi}^{-1}$ (ε_{φ} the energy of the optical phonon), will exclude terms that do not preserve the number of phonons.

Let there be a semiconductor film. When calculating the spectrum, we make the following basic assumptions:

- electrons interact only with optical phonons;
- optical phonons have no dispersion;
- dimensionless coupling constant $a<<1$;
- crystal temperature is low $T<< \varepsilon_{\varphi}$;
- the electron concentration is small, so that all the electrons lie at the bottom of the lower zone of the film ($v = 1$) and the interaction of electrons with each other can be neglected;
- there is no interaction of electrons with impurities and acoustic phonons.

3.2. Solution for the model Hamiltonian

We consider the simplest model in which an electron interacts with one branch of longitudinal optical phonons. For the polarization electron-phonon interaction in a film that does not have a finite radius of action and depends on the pulse transmission module, we have:

$$c_{vv}(\vec{r}) = \frac{1}{V}\sum_{\vec{q}'} c_{vv}(\vec{q}') e^{i\vec{q}'\vec{r}}, \quad (6)$$

where:

$$c_{vv}(\vec{q}') = -\frac{c_0}{q'_\rho} F_{vv}(q'_y),\ \ F_{vv}(q'_y) = \frac{4\pi^2 v^2 (1-\cos q'_y d)}{(4\pi^2 v^2 - q'^2_y d^2) q'_y d} \quad (7)$$

c_0 – amplitude, $c_{vv}(\vec{r})$, $|\vec{q}_1 - \vec{q}_2| = |\vec{q}'|$ – the absolute value of pulse transmission, V – volume, $c_{vv}(\vec{r})$ – the matrix element of the electron-phonon interaction in r – representations, F_{vv} – the matrix element of the electron-phonon interaction in the direction perpendicular to the plane of the film *(y)*, $r=R_0- Q$ – the relative coordinate of the electron and phonon, v – quantum number.
The Schrodinger equation for wave function ψ for a pair of electron+phonon particles in a film, considering the electron-phonon interaction, has the form:

$$\left[-\frac{\hbar^2}{2m^*}\hat{\nabla}^2 + \frac{\hbar^2\pi^2 v^2}{2m^* d^2} + \varepsilon_\varphi + c_{vv}(\vec{r})\right]\psi'(\vec{R}_0\vec{Q}) = \varepsilon\psi'(\vec{R}_0, \vec{Q}) \tag{8}$$

where ε_Φ – phonon energy.

By performing a calibration transformation like [9], we obtain a new equation in which the Hamiltonian does not contain Q. Therefore, the solution can be chosen in the form of:

$$(\vec{R}_0\vec{Q}) = \psi(\vec{r})x(\vec{Q}), \tag{9}$$

where x an arbitrary function.
We will consider $c_{vv}(\vec{r})$ as a disturbance small enough not to mix different levels of the film. Then let's assume

$$\psi(\vec{r}) = \varphi(\vec{r}_\rho)\varphi(\vec{r}_y), \qquad \varphi(r_y) = \sqrt{\frac{2}{d}}\sin\frac{\pi}{d}vr_y \tag{10}$$

where $\varphi(r_y)$ – the wave function of the electron in the direction y. Substituting (9), (10) into (8), multiplying both parts of the resulting equation by $\varphi^*(r_y)$ and integrating by r_y, we find the equation for the wave functions of the bound state:

$$\left[-\frac{\hbar^2}{2m^*}\hat{\nabla}^2 + c_{vv}(\vec{r}_\rho)\right]\varphi(\vec{r}_\rho) = W\varphi(\vec{r}_\rho) \tag{11}$$

where m^* – the effective mass of the electron,

$$W = \varepsilon - \frac{\hbar^2\pi^2 v^2}{2m^* d^2} - \varepsilon_\varphi,$$

$$c_{vv}(r_\rho) = \int_0^d c_{vv}(r)\left|\varphi\left(r_y\right)\right|^2 dr_y = -\sum\frac{4\pi^3 v^4 c_0(\cos q'_y d-1)^2}{r_\rho\left(4\pi^2v^2-{q'_y}^2 d^2\right)^2 {q'_y}^2 d^3} \tag{12}$$

In the polar coordinate system:

$$\hat{\nabla}^2 = \frac{\partial^2}{\partial r_\rho^2} + \frac{1}{r_\rho}\frac{\partial}{\partial r_\rho} + \frac{1}{r_\rho^2}\frac{\partial}{\partial\varphi^2} \tag{13}$$

solving Eq. (11) we get:

$$W_{|m|,n} = -\sum_{q'_y}\frac{8m^*\pi^6 v^8 c_0^2(\cos q'_y d-1)^4}{{q'_y}^4 d^6\hbar^2\left(4\pi^2v^2-{q'_y}^2d^2\right)^4\left(n+m+\frac{1}{2}\right)^2}, \quad n,m = 0,1,2\ldots$$

$$\varphi_{n,m} = \exp(im\varphi)\left(r_\rho\right)^{-\frac{1}{2}}\exp\left(-\gamma\frac{r_\rho}{a}\right)\left(\frac{r_\rho}{a}\right)^{m+\frac{1}{2}}\sum_{k=0}^n a_k\left(\frac{r_\rho}{a}\right)^k, \tag{14}$$

$$-\gamma^2 = \frac{2m^*Wa^2}{\hbar^2}, \qquad a = \sum_{q'_y}\frac{\hbar^2\left(4\pi^2v^2-{q'_y}^2d^2\right){q'_y}^2 d^3}{m^* 4\pi^3 v^4 c_0\left(\cos q'_y d-1\right)^2}$$

Thus the binding energy is negative, i.e., the interaction of an electron and a phonon is an attraction. The spatial separation of an electron and a phonon is determined by the length $\gamma^{-1}a$ and grows with the weakening of the bond. By varying the thickness of the film *(d)*, one can change the binding energy of an electron and a phonon.

As a result thus, in the case of two-dimensional free motion and one-dimensional quantization of the electron and phonon energy spectrum in the electron-phonon system (optical phonons), there are new quasiparticles that are bound states of the electron and phonon.

Let us estimate the value of the binding energy for *GaAs* using the following values of physical quantities: $m^* = m_0 \sim 10^{-27}g$, $d \sim 5\cdot10^{-7}$ cm (film thickness). Then $W_{0,|0|} \sim 5\cdot10^{-18}$ erg, which is beyond the possibilities of observation. However, such an estimate is rough, because the low accuracy of the experiment allows us to determine constants only with an error of several times. For 'successful' crystal, the binding energy can be two orders of magnitude higher, which is already observable.

4. Conclusion

In this paper, we discuss the features of the formation of bound states of phonons and the role of anharmonism for such a formation in a dimensionally quantized system in comparison with bulk sample.

References

[1] Csukanov A V and Kateev I J 2018 *Quantum Electronics* **48(7)** pp 641–47
[2] Wang C T, Wu Z Y, Zhao C L, Ding Z H and Xiao I L 2012 *Chin .Phys. B* **21** 057108
[3] Davydov A C 2010 *Theory of a solid body* (M: Nauka) p 647
[4] Levinson I B and Rashba E I 1972 *Sov. Phys. JETP* **35(4)** 788–93

Distribution of laser radiation in opaque porous media

A A Voznesenskaya, A S Raznoschikov, A F Galkin, E S Oparin, A V Zavitkov, A S Loktev and D S Gusev

Vladimir State University, Gorky Str. 87, 600000 Vladimir, Russia

E-mail: 2obk@bk.ru

Abstract. The work is aimed at studying the interaction of laser radiation and powder material, in particular, the nature of propagation and absorption of laser radiation in a porous medium. Measurements of the nature of propagation and interaction of laser radiation with a powder material were carried out using a low-power laser in the "before reflow" mode. The paper presents the results of measuring the intensity of the passage of a laser beam through layers of powder material made of an aluminum alloy AlSi10Mg with an average fraction of 30 μm of various thicknesses. The results of measuring the heating of the powder material under the action of laser radiation and the dependence of the depth of penetration of the laser beam into the powder material are presented.

1. Introduction

Selective laser melting is a rapidly developing technology for manufacturing parts [1-6]. The main processes in selective laser melting (SLM) are the absorption and scattering of laser radiation, the transfer of laser radiation energy to the granules of the powder material, the melting of the powder material, wetting by the melt and solidification of the melt [7-9]. The totality of the contribution of each of these processes essentially determines the thermal regime of the construction of the part. The transfer of laser radiation energy to the granules of the powder material occurs in a certain volume of the powder. This statement is based on the low bulk density of the material (1.5-2 times less than the density of the base material) and the high reflectivity of the granules. Many models presented on the judgment of the interaction of the laser beam with the outer surface of the granules of the powder material are inconsistent, the results presented in the works are not reproducible in a physical experiment [10-13]. Since the laser beam reflected from the surface of the granules is able to penetrate to a certain depth. Significant thermal resistance of the surrounding gas environment promotes heating of the granules without transferring heat to the surrounding bodies. Since heat transfer between the powder granules is difficult, due to the small area of mutual contact and thermal resistance in the form of a microrelief. The depth of penetration of laser radiation into a layer of a powder material depends on the bulk density, dispersion of particles, the state of the surface of the granules of the powder material [14], and the energy and scanning speed of the laser beam [15]. To have an idea of the absorption of metal powder and the spatial distribution of the absorbed laser radiation in the process of selective laser melting is necessary to determine the optimal parameters and predict possible problems during the assembly of the part. Understanding the nature of the transfer of laser energy to the granules of the powder material underlies the solution to the problem of "mode transfer". This problem occurs when trying to transfer the worked out modes from one installation to another. A widespread approach is to practice the build mode on a small volume machine (laboratory setup) for subsequent reproduction on a machine with a large build chamber volume (process setup). This approach is due to the significantly lower cost of developing the technology, which, in fact, is not justified in view of the significant differences in the characteristics of the focusing optics used to focus the laser beam. This circumstance directly determines the mode of interaction between the laser beam and the powder material, as a result, it is the most important factor that must be taken into account when transferring the mode from one machine to another, at the level of the power parameters and the scanning speed of the laser beam.

The powder is a collection of metal particles, considered as spheres, with dimensions much larger than the laser wavelength (about 1 μm), and with a complex refractive index corresponding to the material and wavelength. The authors of [16] performed complex experimental measurements of normal-hemispherical reflection for various metal powders at wavelengths of 1.06 and 10.6 μm, which are the most common in technological lasers. They found that a layer of powder absorbs significantly better than a flat surface of the same material. Later measurements of normal hemispherical reflection [17,18] and absorption at normal incidence [14] in other scientific groups confirmed the results and expanded the list of studied materials and wavelengths. To evaluate the effect of laser radiation on powder layers, it is necessary to know the spatial distribution of the absorbed energy over depth. The incident radiation can be repeatedly reflected and absorbed by the surface of the powder particles, penetrating deep into the layer of the powder material, which was undertaken within the framework of this work.

The incident laser radiation is absorbed by the granules of the powder material and passes through the powder layer due to multiple reflections through the air gaps [19-21]. Consequently, the absorption of laser energy by a layer of powder is much higher than the absorption by a flat surface of a solid metal of the same composition [16, 20, 22, 23]. The effect of multiple reflections on absorption is more significant for highly reflective metals (such as aluminum and copper) than for moderately absorbing metals (such as iron and titanium); therefore, the dependence of the absorbing capacity on the bulk of the powder and the powder supply system is more pronounced for metals with high reflectivity [19]. The percentage of incident energy absorbed by the substrate or by a previously created solid under the powder layer decreases with an increase in the optical thickness of the powder layer [23] and particle size reduction [24]. In addition to the reflectivity during volumetric heating, an important parameter is the thermal conductivity and melting point of the powder material.

2. Experimental part

In this work, we measured the passage of a laser beam through layers of various thicknesses of powder material of aluminum alloy AlSi10Mg. The study of the granulometric composition of the aluminum powder was carried out using a scanning electron microscope (SEM) and subsequent processing in the Altami Studio software. Granulometric analysis showed that the powder particles are generally spherical in shape, which are shown in Figure 1. The average particle size is 30 μm.

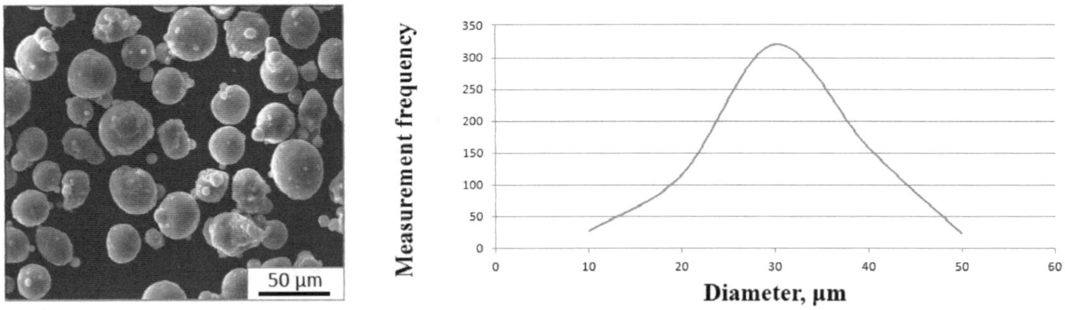

Figure 1. SEM image of the surface of AlSi10Mg powder particles and the results of the particle size distribution of the powder material.

The largest diameter of the particles measured was 49.2 μm. The revealed granule size distribution falls within the range usually used in materials for the manufacture of parts by selective laser melting.

In the work carried out at the stand, the diagram of which is shown in Figure 2, a low-power laser with a wavelength of 650 nm was used. The use of a low-power laser is caused by the need to carry out measurements that exclude phase changes in the material, and as a consequence, significant changes in its physical properties (change in the reflectivity of the resulting surface, change in the thermal conductivity of the melt and heat transfer of the melt bath to surrounding bodies). Precision vertical translator, laser power meter with silicon photodetector. The measurement of the transmitted laser radiation was carried out using the following way: laser radiation with an average power of 1.4 mW with a focusing lens with f = 500 mm simulating the distribution and size of the laser beam close to the distribution of the F-Theta lens of an industrial SLM installation, focused on the surface of the powder material in a spot with a diameter of 100 μm, the photoreceiving area of the silicon photodetector was located under a transparent plate of quartz glass, on the surface of which a layer of powder material was deposited. The vertical transparent plate was moved with a step of 50 μm; at each step of movement, a layer of powder material was added with a leveling knife. To average the result, 10 measurement iterations were performed for each layer thickness of the powder material. The criterion for terminating the experiment was zero laser power.

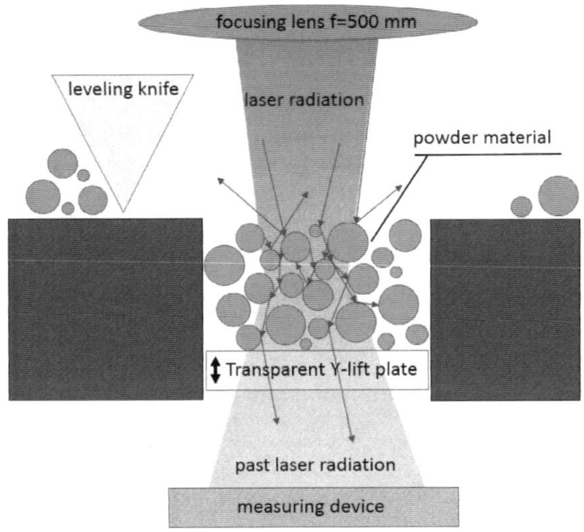

Figure 2. Measuring the transmission of laser radiation in a porous medium.

The second part of the work consisted in measuring the heating of the powder material by laser radiation and the distribution of the temperature front from the heating point. The experiment was carried out on a setup of a similar configuration, except for the use of a zinc sulfide substrate transparent to thermal radiation as a transparent y-lift plate. A thermographic camera operating in the range of 5-8 microns was used as a measuring device. The thickness of the powder layer was varied in the same ranges; the heating of the surface of the powder material was recorded using a thermographic camera. The temperature of the surface of the powder material was measured until the maximum value of the corresponding power of exposure to laser radiation was reached.

3. Research and discussion

The initial thickness of the powder layer was 30 μm, the application of a thinner layer was not carried out, since the average size of the fraction of the used powder was 30 μm. The data were recorded in the protocol for further evaluation of the penetration of laser radiation into the powder layer. The results of measurements of the passage of laser radiation through a layer of powder material are shown in Figure 3. When the thickness of the layer was changed with a step of 50 μm, laser radiation ceased to be recorded by a laser power meter at a layer thickness of 300 μm.

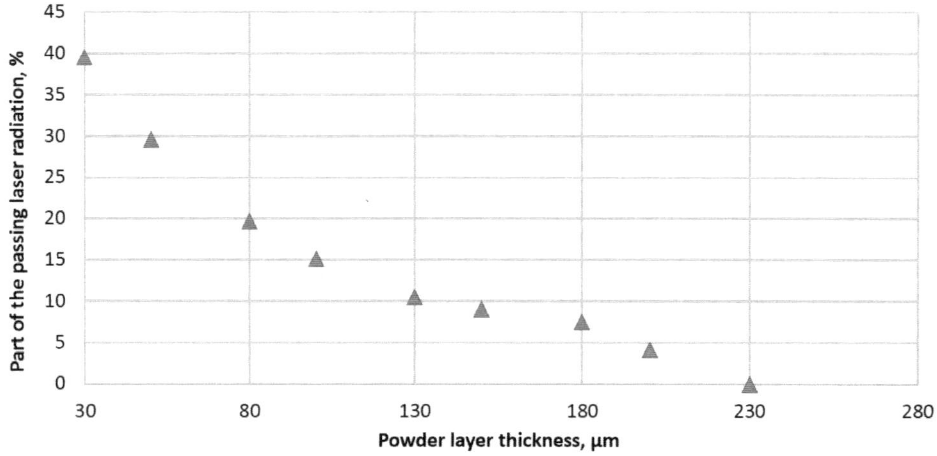

Figure 3. Results of the study of the penetration of laser radiation into the powder layer.

Measurement of the surface temperature of the powder material made it possible to establish that the heating of the powder material did not depend on the thickness of the layer of the powder material (in the range of 30-280 μm). Measurements with a layer thickness of the powder material more than 280 μm were not carried out in order to exclude the phenomenon of thermal conductivity between the granules of the powder material. During measurements, the contrast of the temperature change was about 50-70 μm, which is comparable to the size of the granules of the powder material. That is, only the granules of the powder material are heated, which are exposed to laser radiation and are in direct contact with them. The main mechanism of heat transfer between bodies in the SLM process occurs during the contact of the liquid phase of the material, and largely depends on the wettability of the granule surface by the melt. Figure 4 shows a diagram of the dependence of the surface temperature of the powder material on the power of laser radiation.

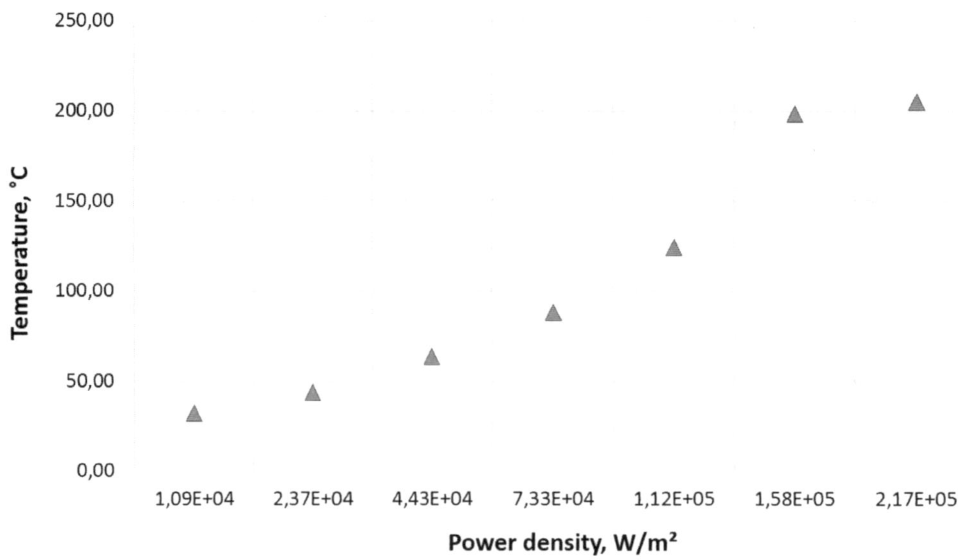

Figure 4. Dependence of the surface temperature of the powder material on the energy of laser radiation.

4. Conclusion

In the study of absorption and transmission of laser radiation through a layer of powder material of various thicknesses, it was found that in the "before melting" mode, laser radiation, due to re-reflection and scattering, penetrates to a considerable depth, much greater than the granule size and layer thickness. The incident laser radiation is almost completely absorbed; only scattering by the surface of the granules of the outer layer is recorded. Measurements using a thermographic camera for heating the powder material made it possible to establish that the layer thickness from 30-300 microns does not affect the temperature of the registered particles.

Acknowledgments

The study was carried out using the equipment of the interregional multispecialty and interdisciplinary center for the collective usage of promising and competitive technologies in the areas of development and application in industry/mechanical engineering of domestic achievements in the field of nanotechnology (Agreement No. 075-15-2021-692 of August 5, 2021).

This work was also partially supported by the framework of the state task of VlSU № 0635-2020-0013.

References

[1] Kruth J P, Froyen L, Van Vaerenbergh J, Mercelis P, Rombouts M and Lauwers B 2004 *J. of mat. proc. techn.* **149 (1-3)** 616–22
[2] Mercelis P and Kruth J 2006 *Rapid Prototyp. J.* **12 (5)** 254–65
[3] Thijs L, Verhaeghe F, Craeghs T, Van Humbeeck J and Kruth J P 2010 *Acta mat.* **58(9)** 3303–12
[4] Louvis E, Fox P and Sutcliffe C J 2011 *J. of Mat. Proc. Tech.* **211(2)** 275–84
[5] Zhang J, Song B, Wei Q, Bourell D and Shi Y 2019 *J. of Mat.Science & Tech.* **35(2)** 270–84
[6] Cai C, Wu X, Liu W, Zhu W, Chen H, Qiu J C D, Sun C-N, Liu J and Shi Y 2020 *J. of Mat. Science & Tech.* **57** 51–64
[7] *Mater.Science Forum* 2020 ed Raznoschikov A S (Stäfa: Trans Tech Publ.Ltd) **989** 845–49
[8] Voznesenskaya A A, Kochuev D A, Raznoschikov A S, Kireev A V and Khorkov K S 2020 *Bulletin of the Russian Academy of Sci.: Phys.* **84(3)** 339–42
[9] Raznoschikov A S, Kochuyev D A, Voznesenskaya A A, Oparin E S and Khorkov K S 2020 *IOP Conf. Ser.: Mat. Sci. and Eng.* **896(1)** 012130
[10] Boley C D, Khairallah S A and Rubenchik A M 2015 *Appl.opt.* **54(9)** 2477–82
[11] Gu D, Yang Y, Xi L, Yang J and Xia M 2019 *Optics & Laser Tech.* **119** 105600
[12] McVey R W, Melnychuk R M, Todd J A and Martukanitz R P 2007 *J.of Laser Appl.* **19(4)** 214–24
[13] Voznesenskaya A, Kochuev D, Zhdanov A, Morozov V and Khorkov K 2018 *J.of Phys.: Conf. Ser.* **1109(1)** 012056
[14] Boley C D, Mitchell S C, Rubenchik A M and Wu S S Q 2016 *Appl. Opt.***55(23)** 6496-6500
[15] Kovalev O B, Gusarov A V and Belyaev V V 2020 *Int. J. of Eng. Sci.* **157** 103378
[16] Tolochko N K, Khlopkov Y V, Mozzharov S E, Ignatiev M B, Laoui T and Titov V I 2000 *Rapid Prototyp. J.* **6(3)** 155–60
[17] Gusarov A V, Bentefour E H, Rombouts M, Froyen L, Glorieuxb C and Kruth J-P 2006 *J. Appl. Phys.* **99** 113528
[18] Zhang D, Wang W, Guo Y, Hu S, Dong D, Schleifenbaum J H and Ziegler S 2019 *J. Mat. Proc. Tech.* **268** 25–36
[19] King W E, Anderson A T, Ferencz R M, Hodge N E, Kamath C, Khairallah S A and Rubenchik A M 2015 *Appl. Phys.Reviews* **2** 041304
[20] Foroozmehr A, Badrossamay M, Foroozmehr E and Golabi S 2016 *Mat. & Des.* **89** 255–63
[21] Gusarov A V and Kruth J-P 2005 *Int. J. of Heat and Mass Transfer* **48** 3423–34
[22] Stacy S C, Zhang X, Pantoya M and Weeks B 2014 *Int. J. of Heat and Mass Transfer* **73** 595–99
[23] Gusarov A V and Smurov I 2010 *Phys. Proc.* **5** 381–94

[24] Zhang B and Coddet C 2016 J. of Manufact. Sci. and Eng., Tran. of the ASME **138** 051001

Fractal model of the surface roughness of laser-induced noble metal nanocluster films

D N Bukharov, A S Abramov, B S Nazarenko, N A Halimov, S P Eyoum Essaka
Vladimir State University, Gorky Str. 87, 600000 Vladimir, Russia

E-mail:buharovdn@gmail.com

Abstract. A fractal approach is proposed for modeling the roughness of nanocluster films obtained by laser synthesis. Models based on the mean-point algorithm, Fourier method and Brownian motion are considered. The parameter connecting the results of modeling and experiment is the Hurst exponent. The similarity of the Hurst exponents for experimental samples and model calculations makes it possible to obtain models with properties that are sufficiently adequate to real samples.

1. Introduction
Nanocluster films of noble metals are widely used in problems of modern biology, nanoelectronics, and opto-photonics [1-3]. Materials based on noble metal nanofilms have new unique properties, largely due to their structural features. Roughness is an important characteristic that can determine new unique properties of a nanostructured surface, for example, hardness, friction, aerodynamic and hydrodynamic resistance, etc. Laser methods for synthesizing samples of nanocluster films make it possible to achieve the required properties in them due to the controlled generation of their relief and roughness. To assess the synthesized relief or to predict it, it is convenient to use the apparatus of mathematical modeling. Below, on the example of films of complexes of silver (Ag) and gold (Au), we will simulate their roughness.

2. Description of experimental samples of nanofilms
The samples of nanocluster structures under consideration were obtained by thermal diffusion deposition on a glass substrate from a previously prepared colloidal solution [4].

Nanofilms of noble metals (Ag, Au, their complexes) on a glass substrate synthesized as a result of laser action were studied using a Quanta 200 3D scanning electron microscope and an Integra-Aura probe nanolaboratory. Analyzing the dynamics of the formation of their relief, it was found that their shape and topology features significantly depend on the number of passes of the radiation beam over the deposition region. For example, for samples of bimetallic Ag / Au nanocluster structures on a glass substrate, obtained at a laser radiation power of 2.5 W and a scanning speed of 1.2 mm / s (Figure 1), it can be noted that after 5 scans a nanocluster film is formed (Figure 1 a) with an average relief height of 14 nm, and already after 10 (Figure 1b), the area of the islands increases significantly, and the average relief height becomes 18 nm.

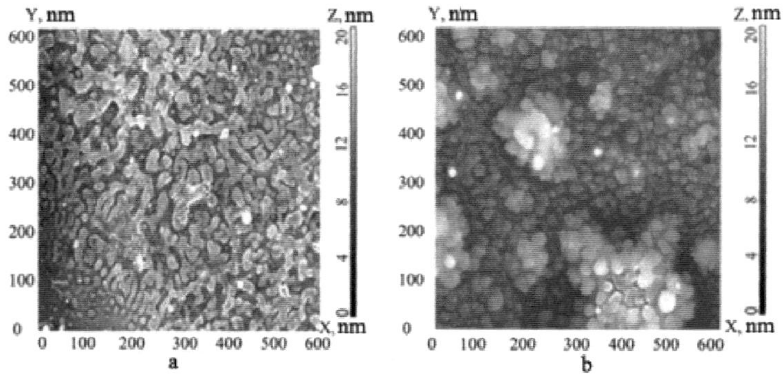

Figure 1. AFM images of Ag / Au bimetallic nanocluster structures after 5 (a), 10 (b) passes.

For the images of the Ag/Au samples, the Hurst exponent (H) was estimated through their fractal (Hausdorf) dimension (D): D=2-H [5]. For the images of the samples from Figure 1, the fractal dimension (D) was estimated based on the boxcounting [6] method and the Hurst exponent (H) was calculated using the above formula.

For the sample from Figure 1, D_1 = 1.8679 and H_1 = 0.1321, for sample 2 – D_2 = 1.8703 and H_2 = 0.1297. The obtained values indicate an increase in the fractal dimension depending on the number of laser beam passes.

3. Description of the surface roughness model

For the calculated Hurst exponent, modeling of rough surfaces was carried out. The surface modeling was carried out on the basis of three popular algorithms that take into account the value of the Hurst exponent: the midpoint displacement method, the Fourier method, and the generalized Brownian motion [7].

Random midpoint displacement method [8] consists of two steps. In the first step an initial square is subdivided into four smaller squares. In the first step we add one vertex into the middle. The added vertex is shifted in z-coordinate direction by random value δ_1. This procedure is recursively repeated for each subsquare, then for every their descendants, and so on.

In order to be resulting surface, the random number δ_i must be generated with Gaussian distribution $\mu=0$, $\sigma=1$ and in the i-th iteration step the variation σ_i have to be modified according to $\sigma_i^2 = \frac{1}{2^{2H(i+1)}}\sigma^2$.

From this equation we can see, that the first iteration has the biggest influence to the resulting shape of the surface and influence of the others decreases.

In the second step we calculate the points on the edges of initial square. We virtually rotate square by 45^0 and calculate the values as in the previous step. The problem is in the cases when the new point has just three neighbors (the encircled points in Figure 2). In this case we calculate the average of three neighbors only. The error produced on the border could be neglected. In the next step we virtually rotate the square back by 45^0 and we recursively apply the first two steps on the four new squares as is mentioned above. This recursive process ends after given number of iteration [9].

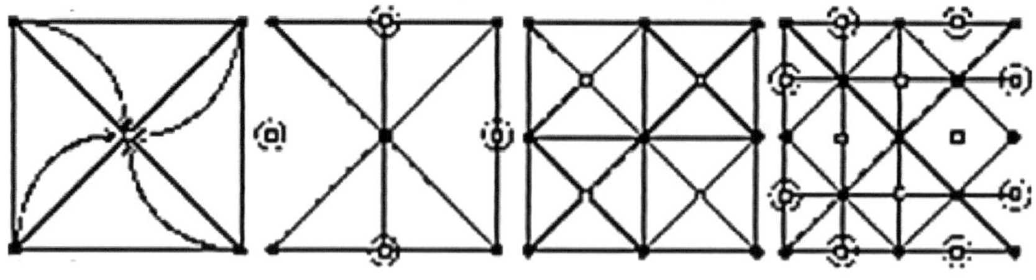

Figure 2. First four steps in random midpoint displacement method.

Figure 3 shows a rough film model based on the random midpoint displacement method with Hurst exponents similar to those in Figure 1. Figure 3a shows a two-dimensional model, Figure 3b shows a three-dimensional image in relative units.

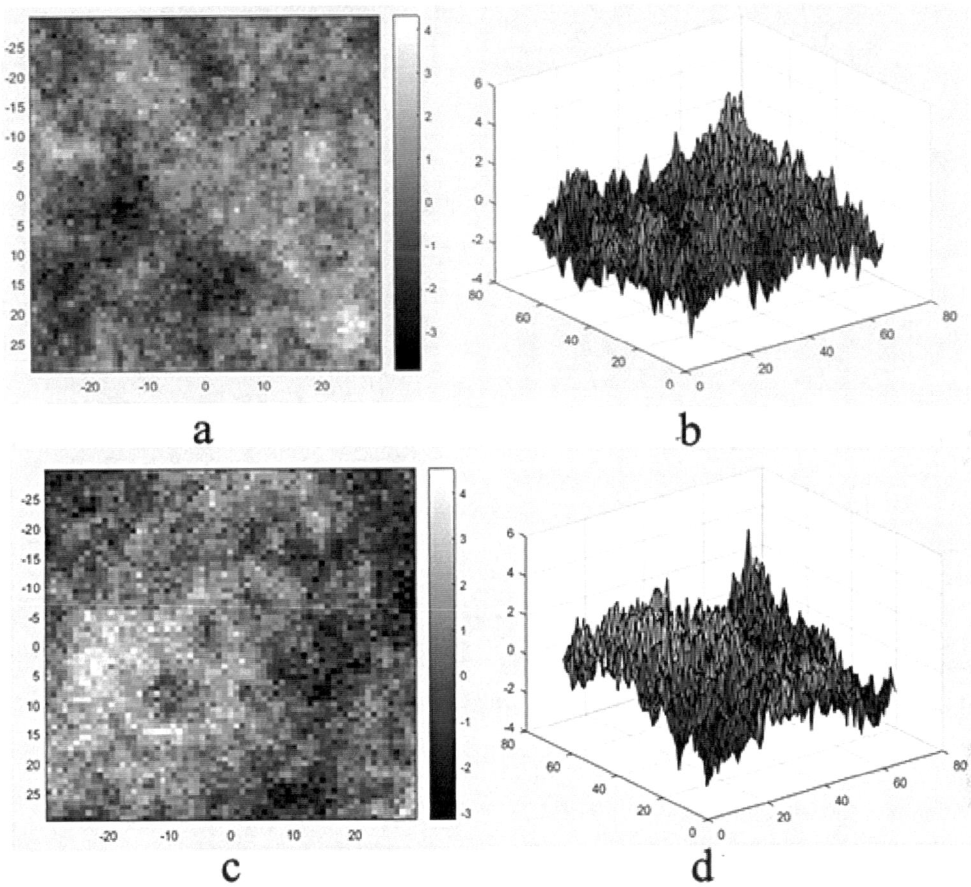

Figure 3. Rough film model based on the random midpoint displacement method: H=0.1321, 2d (a), 3d(b), H=0.1279, 2d (c), 3d(d).

Fourier method (also called Spectral synthesis) is another method to generate surface taking into account Hurst exponent [10]. First step we generate spectral density function $S(\boldsymbol{q}) = \left(\sqrt{q_1^2 + q_2^2}\right)^{-2(H+1)}$,where $\mathbf{q} = (q_1,\ q_2)$ is the frequency in Fourier space, computing the complex coefficients ($\eta(\mathbf{q_1,q_2})$) of inverse Fourier transformation: $\eta(q) = c(\boldsymbol{q})\exp\big(2\pi\varphi(\boldsymbol{q})\big)$,where $\mathrm{c}(\mathbf{q})$ the magnitude and $\phi(\mathbf{q})$ the phase. Then we calculate the inverse Fourier transformation in two dimensions at coordinate x_1 and x_2:

$$h(x_1, x_2) = \sum_{q_1=0}^{N-1} \sum_{q_2=0}^{N-1} \eta(q_1, q_2) \exp(-2\pi i(q_1 x_1 + q_2 x_2). \text{ [11]}$$

Figure 4 shows a rough film model based on the Spectral synthesis method with Hurst exponents similar to those in Figure 1. Figure 4a shows a two-dimensional model, Figure 3b shows a three-dimensional image in relative units.

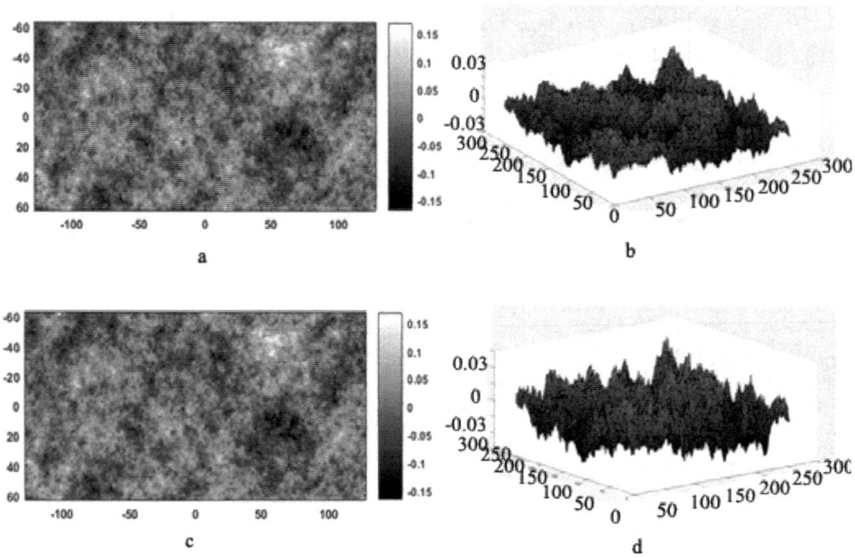

Figure 4. Rough film model based on the random midpoint displacement method: H=0.1321, 2d (a), 3d (b), H=0.1279, 2d (c), 3d(d).

Generalized Brownian motion is a model for describing motion some hypothetical particle that makes a chaotic movement in space. Particle motion is considered as the result of a large number of small random displacements. In contrast to the classical Brownian motion in the generalized model, individual steps movements are not completely random and independent, but are determined by the history movements in all previous moments. For the displacement dependence of the generalized Brownian particles along one of the spatial coordinates from time can be written as follows ratio: $B_H(t) = \frac{1}{\Gamma(H+1/2\_}\int_{-\infty}^{t} K(t-t')dB(t')$, where Γ is the gamma function, $dB(t')$ - independent random movements an ordinary Brownian particle at times t'; function $K(t-t')$, which is the kernel integral expression, is responsible for the memory of the system and is defined as follows ratio:

$$K(t-t') = \begin{cases} (t-t')^{H-1/2}, 0 \le t' \le t \\ (t-t')^{H-1/2} - (-t')^{H-1/2}, t' < 0 \end{cases}.$$

Figure 5 shows a rough film model based on the Generalized Brownian motion method with Hurst exponents similar to those in Figure 1. Figure 5a shows a two-dimensional model, Figure 3b shows a three-dimensional image in relative units.

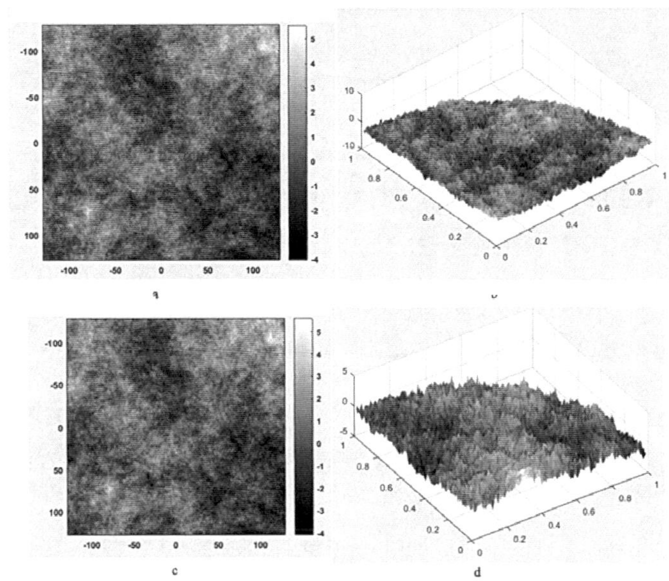

Figure 5. Rough film model based on the Generalized Brownian motion method: H=0.1321, 2d (a), 3d (b), H=0.1279, 2d (c), 3d(d).

4. Conclusion

The proposed approach can be useful for the problems of synthesizing materials with the required properties. The considered models are of a random nature, but admit the assessment of the adequacy by experimental results in the sense of comparing the Hurst exponents.

Acknowledgments

The study was carried out using the equipment of the interregional multispecialty and interdisciplinary center for the collective usage of promising and competitive technologies in the areas of development and application in industry/mechanical engineering of domestic achievements in the field of nanotechnology (Agreement No. 075-15-2021-692 of August 5, 2021).

This work was carried out as part of the RFBR project №20-02-00515 and was also partially supported by the framework of the state task of VlSU № 0635-2020-0013.

References

[1]. Cobley C M, Chen J, Cho E C, Wang L V and Xia Y 2011 *Chem. Soc. Rev.* **40** 44–56
[2]. Sun J and Jin Y 2014 *J. Mater. Chem. C* **2** 8000–11
[3]. *Nanostructured Materials and Nanotechnology* 2002 ed Nalwa H S (Gulf Professional Publishing) p 834
[4]. Arakelyan S M, Kucherik A O, Prokoshev V G, Rau V G and Sergeev AG 2015 *Introduction to femtosecond nanophotonics, Fundamental principles and methods of laser diagnostics and control of nanostructured materials* (Moscow: Logos) p 744
[5]. Uthayakumar P and J Gopalakrishnan 2018 *Eng. Technol. Appl. Sci. Res.* **6** 1210–15
[6]. Gonzato G 1998 *Comput. Geosci.* **24** 95–100
[7]. Barabash T K and Maslovskaya A G 2010 *Bull. Amur State. Univ. Series: natural and economic sciences* **49** 31–38
[8]. Fournier A, Fussel D and Carpenter L 1982 *Commun. ACM* **25** 371-84
[9]. Marák I 2000 *CESCG 97-99 Selected Papers* (Vienna: Osterreichische Computer Gesellschaft)
[10]. Vaseghi S V 2000 *Advanced Digital Signal Processing and Noise Reduction* (Hoboken: John Wiley & Sons, Inc.) p 514
[11]. De Castro C P, Luković M, Andrade R F S and Herrmann H J 2017 *Sci Rep* **7** 1961

The spectral characteristics of plasma-erosion torch on the surface of carbon-bearing materials

S M Arakelyan, A A Gadalov, A F Galkin, S V Zhirnova and E L Shamanskaya

Vladimir State University, 87 Gorky Street, Vladimir 600000, Russia

E-mail: svetlanna_vik@mail.ru

Abstract. The paper obtains spectrograms of the plasma-erosion torch on the surface of the glass carbon in the air and in the vacuum, which allow us to judge the presence of certain particles and their number in the torch depending on the parameters of laser radiation. Experiments have shown that the composition of the plasma-erosion torch in air and in vacuum has a complex structure with different particles, depends on the parameters of laser radiation, which can affect the characteristics and modes of deposition of carbon films.

1. Introduction

Plasma technologies are a group of technologies for obtaining and processing materials using heating of initial products in a plasma jet or transferring them to the plasma state. Plasma technologies are a group of technologies for obtaining and processing materials using heating of initial products in a plasma jet or transferring them to the plasma state.

One of the most common ways to modify the surface using plasma technologies is the application of thin-film coatings by laser-plasma method [1]. It has several advantages over other methods: first, the possibility of obtaining plasma from any substance, and secondly, when applying nanostructured films in a vacuum, laser radiation does not introduce additional impurities into the precipitated coating. In addition, the laser-plasma method allows you to control the composition of the precipituted coatings. To obtain the required characteristics of nanofilms, it is necessary to study the features of the destruction of the laser target material in a vacuum, to determine the number of ionized and excited atoms in the plasma, the dynamics of the parameters of the erosion laser torch and its spatial characteristics. The composition of the plasma-erosion torch affects the characteristics of thin films, allows you to control the composition of the preciphered coatings [2-3].

The aim of the work is to study the composition of the plasma-erosion torch of a glass-carbon target during laser ablation in a vacuum using spectroscopy of laser-induced plasma with variation in the parameters of laser radiation. The main advantage of vacuum plasma treatment is a good pre-vacuum cleaning of the treated material from moisture, fats, oils, oxides, embedded gases and other substances. The resulting coatings have properties that are unattainable with other methods of formation [1].

2. Experimental installation

This paper presents the results of experimental studies of the emission spectra of the plasma-erosion torch generated by laser pulses of various parameters on the surface of the glass-carbon target in a vacuum. The scheme of the experimental installation is shown in Figure 1. In the figure, the numbers indicate: 1 – turbopump unit HiClube 80 Eco; 2 – pulse-periodic laser CLW50 (femtosecond laser system TETA-10); 3 – vacuum chamber; 4 – a sample of glass carbon fiber; 5 – spectrometer ASP-75.

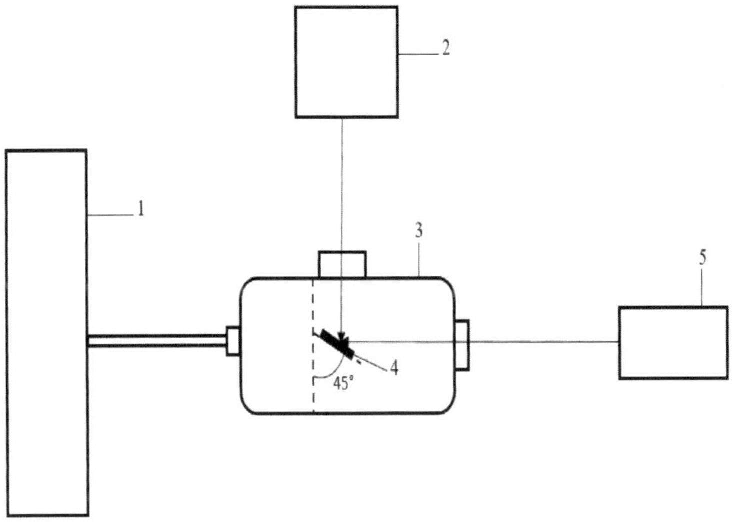

Figure 1. Scheme of experimental installation.

The pulse-periodic laser CLW-50 was used as a power laser. The radiation of the Nd:YAG laser with a working wavelength of 1064 nm, coming out of the laser emitter through a fiber-optic cable, is fed into the optical-mechanical system to form a spot of the desired diameter in the impact zone. Radiation through the camera window hits the surface of the sample. As an alternative laser, the FEMtosecond laser system TETA-10 with a central wavelength of radiation of 1029 ± 2 nm was used. To determine the composition of the plasma-erosion torch, the ASP-75 spectrometer was used, which is mounted against the second window of the chamber. A vacuum is created in the chamber with the help of the turbopump unit of the HiClube 80 Eco Turbopump Unit.

Figure 2 shows the appearance of the experimental installation. In the figure, the numbers indicate: 1 – spectrometer ASP-75; 2 – vacuum chamber in which the glass carbon sample is located; 3 – femtosecond laser system TETA-10; 4 – turbopump unit HiClube 80 Eco.

Figure 2. Appearance of the experimental installation.

The energy in the pulse (E) varied from 1 J to 2 J, the pulse repetition rate (f) varied from 45 Hz to 50 Hz, the pulse duration (τ) varied from 0.5 ms to 2 ms. The diameter of the laser beam was 0.4 mm. As an alternative laser, the femtosecond laser system TETA-10 with a central radiation wavelength of 1029 ± 2 nm, with an energy in a pulse of 140 μJ, with a pulse duration of 280 fs, was used; pulse repetition frequency varied from 10 to 20 kHz. In the chamber in which the sample of glass carbon was placed, a vacuum was created using the HiClube 80 Eco turbopump unit. The ASP-75 spectrometer was used to determine the composition of the plasma-erosion torch.

Figure 3 shows an image of plasma-erosion on the surface of the target glass carbon in the air. The numbers indicate: 1 – glow near the surface of the target; 2 – glow at a distance of ~3 cm from the target surface; 3 – a sample of glass carbon. Laser parameters: E = 2 J; f = 45 Hz; τ = 2 ms.

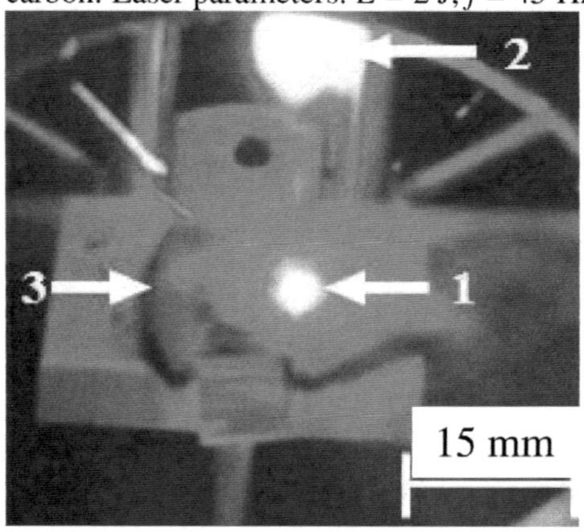

Figure 3. Spatial glow of the erosion laser torch on the surface of the target of glass carbon fiber.

3. Experimental results

During the experiment, several spectra of the plasma-erosion torch were obtained at various parameters of laser radiation acting on the surface of glass carbon in air and in vacuum.

Figure 4 shows the emission spectra of the plasma-erosion torch obtained at the following parameters of laser radiation in the air: a) pulse-periodic laser CLW-50 with E = 2.2 J, f = 35 Hz, τ = 2 ms; b) femtosecond laser system TETA-10 with E = 140 μJ, f = 20 kHz, τ = 280 fs.

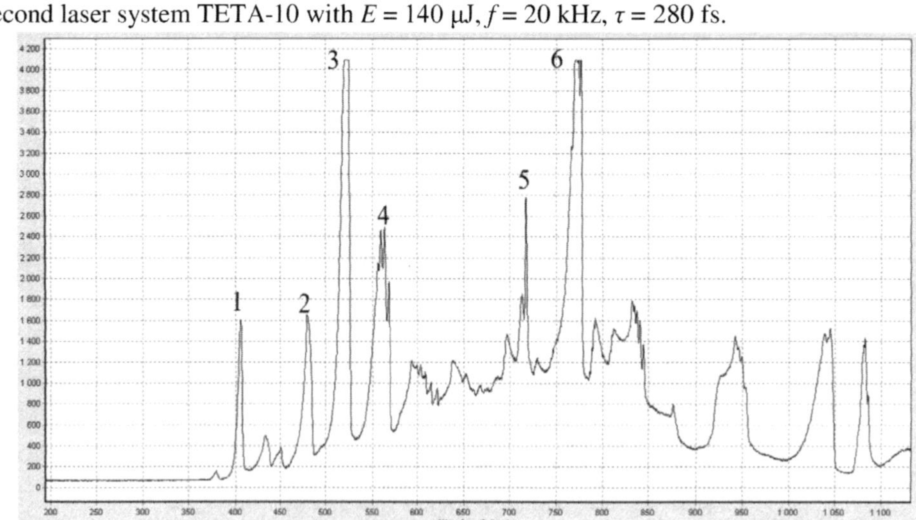

(a)

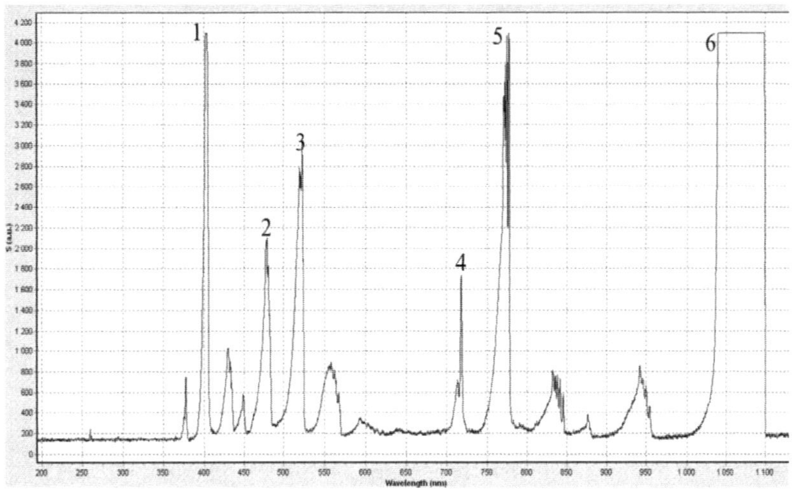

(b)

Figure 4. Emission spectra of plasma-erosion torch in the air

The graph shows the characteristic maxima corresponding to the radiation intensity of the particles that make up the plasma-erosion torch. The numbers indicate the maximum wavelengths that correspond to the radiation of the following excited ions. Analysis of the luminescence spectra of the plasma-erosion torch in the air showed that the following types of ionized atoms have in the torch: C^{1+}, C^{3+}, Ar^{1+}, Ar^{2+}, N^{2+}, N^{3+} [4].

Figure 5 shows the emission spectra of the plasma-erosion torch obtained at the following parameters of laser radiation in vacuum: a) pulse-periodic laser CLW-50 with $E = 2.2$ J, $f = 35$ Hz, $\tau = 0.5$ ms; b) femtosecond laser system TETA-10 with $E = 140$ μJ, $f = 10$ kHz, $\tau = 280$ fs.

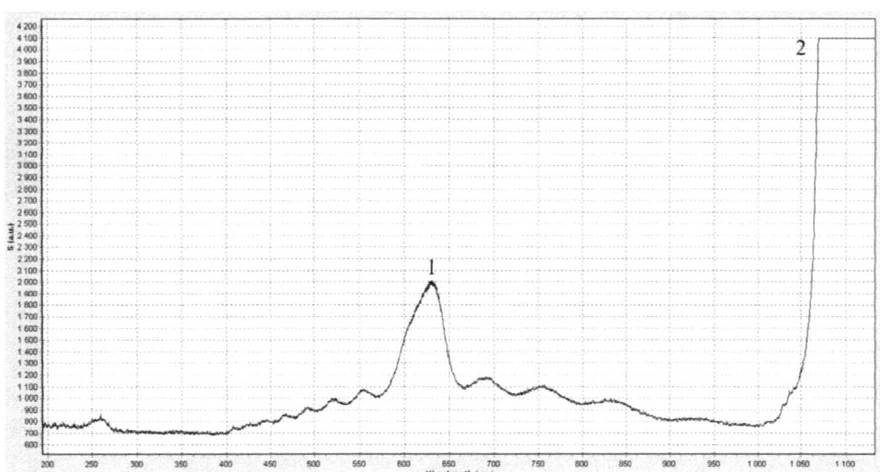

(a)

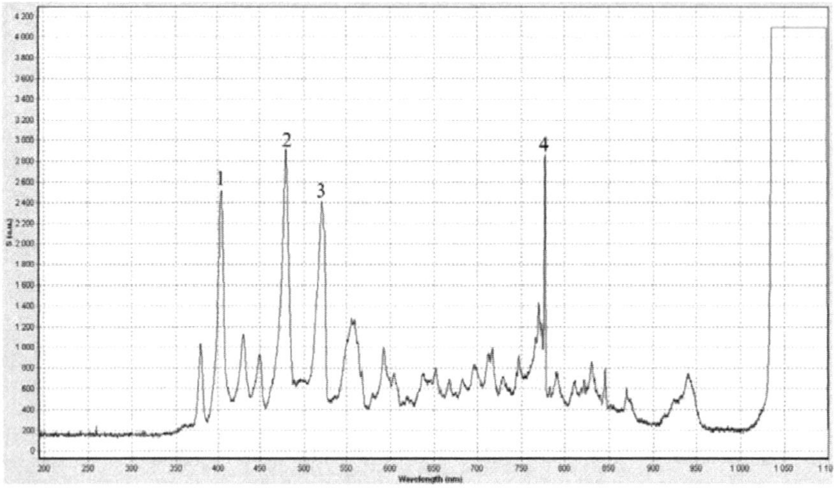

(b)

Figure 5. Emission spectra of plasma-erosion torch in vacuum

Analysis of spectra obtained in vacuum when exposed to laser radiation with a millisecond pulse duration, the plasma-erosion torch has one type of ionized carbon atom C^{1+}. When exposed to the surface of glass carbon laser radiation with femtosecond pulse duration in a vacuum, the torch contains two types of ionized carbon atoms C^{1+}, C^{2+}. With an increase in the energy of laser radiation in the pulse, the concentration of carbon ions increases. This is evidenced by the increase in the relative intensity of these spectral lines.

Analysis of the luminescence spectra of the plasma-erosion torch obtained in the air and in vacuum showed that in the air as part of the plasma-erosion torch there are ionized atoms of various elements, which complicates the process of obtaining pure carbon films.

4. Conclusion

The paper obtains spectrograms of a plasma-erosion torch near the surface of glass carbon in air and in vacuum, which allow us to judge the presence of certain particles and their number in the torch depending on the parameters of laser radiation.

Thus, experiments have shown that the composition of the plasma-erosion torch in air and in vacuum has a complex structure with different particles, depends on the parameters of laser radiation, which can affect the characteristics and modes of deposition of carbon films.

Acknowledgments

This work supported by the RFBR project № 20-02-00515 and also partially supported by the framework of the state task of VlSU № 0635-2020-0013.

The study was carried out using the equipment of the interregional multispecialty and interdisciplinary center for the collective usage of promising and competitive technologies in the areas of development and application in industry/mechanical engineering of domestic achievements in the field of nanotechnology (Agreement No. 075-15-2021-692 of August 5, 2021).

References

[1]. Krivobokov V, Sochugov N and Solovyov A 2011 *Plasma coatings (properties and applications)* (Tomsk: Tomsk Polytech. Univer. Press) p 137

[2]. Goncharov V, Puzyrev M and Stupakevich V 2016 *Vestnik BSU* **1**(1) 79–83

[3]. Agarkova E 2015 *Mat. of the VII International Student Scientific Conference* (15.02-31.03.2015 Moscow RU)

[4]. Saidel A 1965 *Fundamentals of spectral analysis* (Moscow: Sci. Press) p 324

AUTHOR INDEX

Institute of Physics
Dirac House, Temple Back
Bristol BS1 6BE UK

ISSN: 1742-6588
ISBN 978-1-7138-6037-2

CURRAN ASSOCIATES INC.
proceedings.com